SEM Petrology Atlas

by Joann E. Welton

Chevron Oil Field Research Company

Methods in Exploration Series
Published by
The American Association of Petroleum Geologists
Tulsa, Oklahoma 74101, USA

Published by
The American Association of Petroleum Geologists
Tulsa, Oklahoma 74101, USA

for AAPG:
Editor: Richard Steinmetz
Science Director: E.A. Beaumont
Project Editor: R.L. Hart

Welton, Joann E., 1950-
    SEM petrology atlas.

    (The AAPG methods in exploration series; no. 4)
    Bibliography: p.
    1. Mineralogy, Determinative--Atlases.   2. Scanning
electron microscope.   I. Title.   II. Title: S.E.M. petrology
atlas.   III. Series.
    QE369.M5W45 1984   549'.1145   84-6225
    ISBN 0-89181-653-4

Table of Contents

Publisher's Note:

The American Association of Petroleum Geologists gratefully acknowledges the management and personnel of Chevron Oil Field Research Company, for their contribution of this manual to the profession. This comparative atlas was developed to assist geologists working for Chevron in their everyday work of examining sedimentary minerals in exploration and reservoir development. It was given to AAPG for publication so that others could share in its usefulness.

The use of scanning electron microscopy, X-ray diffraction, and energy dispersive X-ray has greatly increased over the past few years as the equipment is now available at more levels within industry and academia. Certainly geologists are able to have their samples sent away to a service lab for analysis. In this spirit, we've published this book to assist the generation of geologists still in school to be exposed to these uses; and we've published this book to assist the generation of geologists at work in industry to become acquainted with this valuable tool for exploration and development.

AAPG Publications
Tulsa, Oklahoma

Acknowledgments

This book is a slightly modified version of a research report compiled at Chevron Oil Field Research Company, La Habra, California. I would like to express my appreciation to the management of Chevron Oil Field Research Company, in particular J.R. Baroffio, F.L. Campbell, and L.C. Bonham for their encouragement and permission to publish this atlas. Special thanks to R.L. Burtner who suggested I compile this atlas and to R.L. Burtner and M.N. Bass for their many hours of careful review which greatly improved the final text. I would also like to thank my other colleagues at Chevron who reviewed this atlas and provided helpful suggestions during various phases of this project: H.M. Beck, A.B. Carpenter, E.W. Christensen, J.R. Frank, E.L. King, D.R. Kosiur, A. Levison, C.A. Meyer, D.W. Richards, G.W. Smith, M.A. Warner, and B.J. Welton.

Special thanks to J.M. Peterson (Keplinger and Associates) who reviewed the text and S.S. Ali (Gulf) who reviewed the bibliography for the AAPG. R.C. Surdam (University of Wyoming) and R.A. Sheppard (USGS) kindly provided formation and age information on the zeolite samples.

Finally, I am grateful to the many people who helped in preparation of the atlas, in particular, N.E. Breen, J.A.B. Quinn, V.E. Welsh, T.N. Bube, D.K. Kitazumi, J.C. Keeser, C.F. Everett, S.K. Elmassian (cover illustration), and V.K. Salvi of Chevron, and to the staff of the AAPG, in particular E.A. Beaumont and R.L. Hart, for their patience in the midst of many revisions and without whose help this atlas would not have been possible.

J.E. Welton
Chevron Oil Field Research Company
La Habra, California

In the last few years, our need to answer complex exploration and production questions has led to the use of increasingly sophisticated analytical equipment. Today, the scanning electron microscope (SEM) and energy dispersive X-ray (EDX) systems are being successfully applied to a wide variety of petroleum exploration and production problems. These include: (1) identification of plant and animal microfossils (for age and environmental interpretations); (2) evaluation of reservoir quality through diagenetic studies; and (3) the investigation of production problems, such as the effect of clay minerals, steamfloods, and chemical treatments on drilling equipment, gravel packs, and the reservoir itself.

Although the use and application of the SEM has steadily increased, the amount of reference material available to aid in SEM mineral identification has severely lagged behind. Some textbooks are available which give excellent descriptions of basic SEM theory (Postek et al, 1980; Wells, 1974), but these books are not written specifically for geologists, so are limited as a geologically oriented SEM work. Papers dealing with the identification of authigenic clay minerals (Wilson and Pittman, 1977) and zeolites (Mumpton and Ormsby, 1976) are an excellent beginning, but no *comprehensive* guide to mineral identification by SEM has been available.

The purpose of this atlas is to provide SEM users (geologists, engineers, geochemists, and technicians) with a beginning guide to SEM mineral identification and interpretation. This atlas by no means contains a complete representation of all common minerals, but rather includes a wide variety of minerals commonly found in clastic reservoir rocks.

Why SEM Analysis?

Since the 1800s, thin-section analysis of rocks using a polarizing or petrographic microscope has been a traditional tool of the geologist. With the petrographic microscope, geologists are able to examine a two-dimensional cross section through a rock, estimate the bulk mineral composition, and make important observations regarding grain fabric and texture. However, the actual three-dimensional grain relationships and details of the intergranular pore structure were always beyond our reach.

With the introduction of the SEM and EDX systems, geologists are now able to go one step beyond thin section analysis — to look down into the pores, identify the smallest minerals, and examine the distribution of these minerals within the pores. Other advantages of the SEM over optical petrography are ease of sample preparation, greater depth of field and resolution, and a significantly higher magnification range (most SEM analysis of rocks involves magnifications between 10× to 20,000×). In addition, less training is required to interpret an SEM micrograph. When examining an SEM micrograph for the first time, the major problem is one of scale. But, with minimal training and experience, the user can soon identify minerals and textures previously observed only in thin section.

This is not to say that the SEM replaces thin section analysis; instead, the SEM complements thin section analysis by providing a different type of information which — when used in combination with other techniques — provides important new information to help characterize rocks.

Format

The most reliable way to identify minerals through the SEM is to compare their characteristic morphologies (such as those shown in this atlas) with the elemental compositions determined by the EDX system. This atlas includes both SEM micrographs and EDX spectra for most of the common minerals found in sedimentary rocks. Identifications were verified by X-ray diffraction analysis when possible. As most geologists are trained to examine and interpret petrographic thin sections, I have included some thin section examples showing similar features. Hopefully, this will ease the transition from thin section to SEM analysis and emphasize the complementary aspects of the two techniques.

All of the minerals illustrated in this atlas are grouped according to species (e.g. silicates, carbonates); the silicates are further subdivided into important mineral groups (silica, feldspars, clays). Most plates consist of a series of three micrographs arranged in order of increasing magnification and accompanied by a brief interpretive description (Figure 1 shows a key to the alphanumerics at the bottom of each micrograph). The SEM micrographs were taken using an ETEC Autoscan U-1 scanning electron microscope (20 KV). The EDX spectrum for that mineral is located on the opposite page (the white circle in the SEM micrograph indicates the location of the EDX analysis). The EDX spectra were obtained using a KEVEX 5000μx energy dispersive fluorescence X-ray system. In a few cases (hornblende, magnetite, etc.), only an EDX spectrum and one SEM micrograph are included. These individual spectra provide additional examples of important common minerals for comparison.

Most of the examples were extracted from Chevron Oil Field Research Company (COFRC) technical service projects. In order to more thoroughly document and characterize certain important mineral groups, examples are also included from the API Reference Clay Minerals suite, X-ray diffraction standards, and a zeolite reference suite (purchased from Minerals Re-

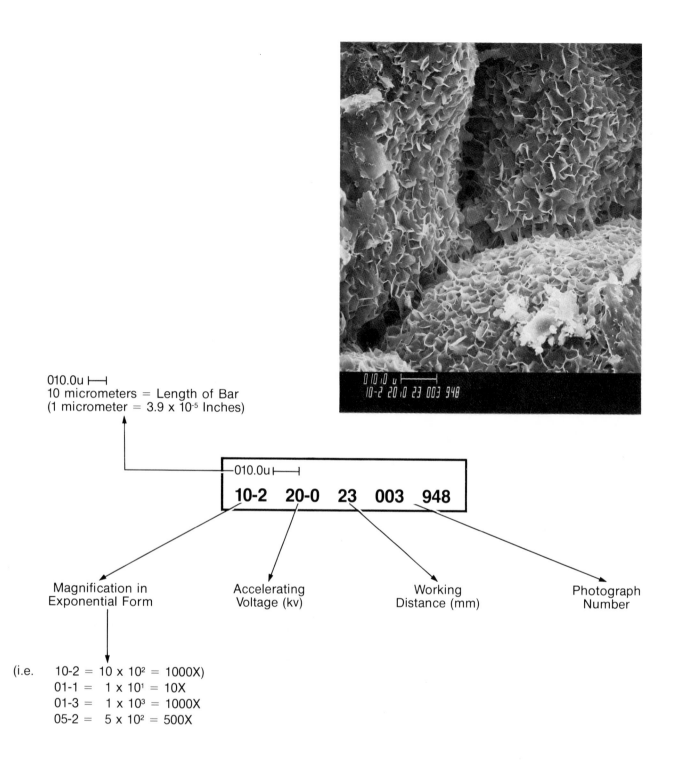

010.0u ⊢──┤
10 micrometers = Length of Bar
(1 micrometer = 3.9 x 10^{-5} Inches)

┌─ 010.0u ⊢──┤
│ **10-2   20-0   23   003   948**
└────────────────────────────────

Magnification in
Exponential Form

Accelerating
Voltage (kv)

Working
Distance (mm)

Photograph
Number

(i.e.   10-2 = 10 x 10^2 = 1000X)
01-1 = 1 x 10^1 = 10X
01-3 = 1 x 10^3 = 1000X
05-2 = 5 x 10^2 = 500X

Figure 1 — SEM Alphanumeric Key

search, Clarkson, New York).

A brief introductory description of sample preparation and SEM/EDX theory follows this introduction. Following the text is a list of selected references emphasizing the geologic application of the SEM, an X-ray Energy Table, and a glossary of geologic terms used. All definitions were compiled from the AGI· Glossary of Geology (1972). All chemical formulas were derived from Deer, Howie and Zussman (1966).

Abbreviations used in this atlas are: TS = thin section; PL = plane light; XN = crossed nicols; CL = cathodoluminescence; SEM = scanning electron microscope; EDX = energy dispersive X-ray; WDX = wavelength dispersive X-ray; XRD = X-ray diffraction, eV = electron volt; KeV = Kilo electron volt; KV = kilovolts.

Sample Preparation

SEM analysis can be done on a wide variety of materials (for example, core and sidewall samples, drill cuttings, thin sections, corroded tubing). The major requirement is that the sample be small enough to fit into the SEM sample chamber.

Rock samples submitted for SEM analysis should be large enough that a fresh surface, uncontaminated by drilling fluids, can be obtained. For log and petrophysical comparisons using core, it is desirable that the SEM, thin section, porosity, permeability, cation exchange capacity (CEC), and X-ray diffraction samples be taken at the same depth. Usually a 1 by 2 in. (25 by 50 mm) core plug will provide sufficient material for all of these analyses. To minimize column contamination, oil-coated samples can be cleaned in a soxhlet extractor with solvents such as a 20/80 chloroform-acetone azeotrope for 24 to 48 hours.

The SEM sample is obtained by gently breaking the rock or core plug with a small rock-chopper or X-acto knife. Be careful not to introduce artifacts by scraping the knife across the surface to be examined. Optimal size for the final sample is generally around 5 by 10 by 10 mm. Any fine debris on the surface can usually be dislodged with a Freon duster. For best results, samples should be handled with disposable gloves, tongs, tweezers, etc., because skin oil from fingers will outgas in the SEM vacuum system, degrading the SEM image.

The cut sample is attached to a SEM specimen plug with epoxy or Silpaste and dried overnight in a low-temperature drying oven. A thin line of Silpaint is added to provide an electrical ground from the sample to the plug. The sample is then coated with a conductive metal, such as carbon, gold, or palladium in either a sputter or evaporative coater. This coating is required to obtain a clear image of an insulating material (such as a rock sample), but is so thin (200 Å) that it does not hinder the identification of specific minerals.

We found that for porous sandstones, a combined coating of carbon plus gold (or palladium) gives the best results. Most of the samples shown in this atlas were coated with carbon and gold in a Kinney evaporative coater. After coating, the sample is ready for SEM analysis.

How The SEM Works

The scanning electron microscope consists of an electron optics column and an electronics console (Figure 2). The coated SEM sample is placed in the sample chamber, in the electron optics column and evacuated to high vacuum (approximately 2×10^{-6} torr).

Instead of using light, as in the petrographic microscope, the SEM image is formed by an internally generated electron beam. This beam is created by heating a "hairpin" tungsten filament (Figure 2) in the electron gun until the filament emits electrons. The electrons are accelerated through the column by a 5- to 30-KV accelerating voltage, demagnified and focused through a series of electromagnetic lenses into a finely-focused beam, which bombards the sample. Final diameter of the beam is typically 100 angstroms ($1 Å = 10^{-8}$ cm) in most commercial SEM's. Additional components include a stigmator for controlling the shape of the beam and apertures to minimize lens defects (aberrations), which in light microscopy severely limit resolution.

It is the interaction of the primary electron beam with the sample which produces various forms of radiation, such as secondary electrons, characteristic X-rays, auger electrons, backscatter electrons, and "bremsstrahlung" (continuous or background) X-rays. As all of these reactions occur simultaneously, it is possible to both observe and analyze the elemental composition of an isolated mineral in a matter of seconds. In geologic analysis, we primarily use the secondary electrons (SEM micrograph) and the characteristic X-rays (EDX spectrum). The remainder of the radiation is dissipated into heat or lost in the walls of the SEM sample chamber.

The SEM Micrograph

The three-dimensional topographic image (SEM micrograph) is formed by collecting the secondary electrons generated by the primary beam. These are low-energy electrons, so only those formed near the surface (50 to 500 Å deep for insulating materials such as rocks) are able to escape (Wells, 1974). As the electron beam traverses the sample, the secondary electrons emitted are collected by a secondary electron detector mounted in the SEM sample chamber and processed by the electronics console into the familiar SEM im-

Scanning Electron Microscope
and
Energy Dispersive X-Ray Spectrometer

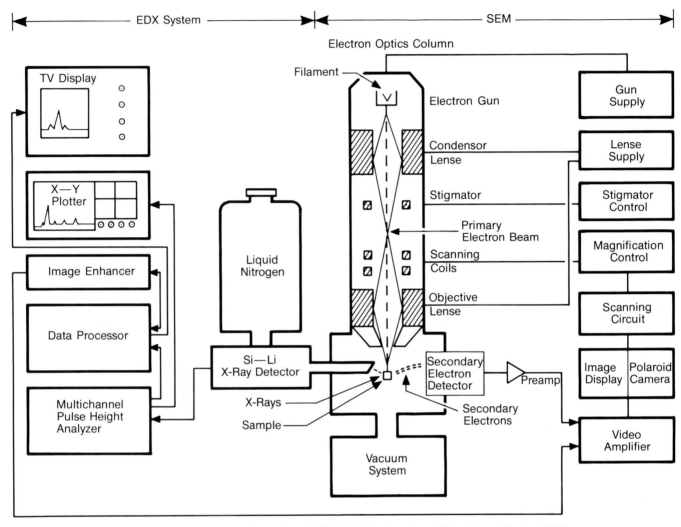

Figure 2   Schematic Showing SEM/EDX System (modified from Beck, 1977)

age. This image is either displayed on a TV screen or photographed with an attached Polaroid camera.

Elemental Analysis By SEM

Elemental analysis of a sample is obtained by collecting the characteristic X-rays generated as the electron beam scans the sample. The X-ray detector is mounted adjacent to the secondary electron detector (Figure 2). Each element in the sample produces X-rays with characteristic energies and wavelengths. These X-rays can be analyzed using an energy sensitive Si(Li) detector in an energy dispersive system

(EDX) or by dispersing the X-rays according to wavelength using the crystal detector of a wavelength dispersive system (WDX). In general, the EDX system is used to obtain rapid analysis of elements above atomic number (Z) = 11 (sodium), whereas the WDX system yields precise quantitative analyses, including light or trace elements.

The major differences between the two systems are:

1. The EDX system yields quick, low-cost analysis of all elements in the sample simultaneously; the WDX system analyzes only one element at a time, making analysis slower and more expensive.

2. Resolution is poor in the EDX system (150 eV/channel), but excellent in the WDX system (2 to 20 eV/channel).

3. EDX analysis can be done on either rough-cut or polished thin sections; WDX analysis can only be done on polished thin sections.

4. EDX analysis yields semiquantitative data (if polished thin sections are used, the data is more precise); the WDX system yields precise quantitative data.

In this atlas X-ray elemental analysis will deal only with EDX systems. For additional information on WDX analysis, see Smith (1976) and Postek and others (1980).

EDX Analysis of Minerals

In EDX analysis, the primary electron beam in the SEM ionizes the atoms of the mineral being analyzed by exciting and ejecting electrons in the inner shells of the atoms. To regain stability, electrons from the outer shells replace the inner shell vacancies (Figure 3). These transitions from outer to inner shell release specific amounts of energy, in the form of X-rays. The energy of each X-ray is determined by the energy difference between the electron shells involved, differences in the electron spin, and the number of protons in the nucleus. Only the strongest of these transitions are detected by the EDX system. The K X-rays represent the strongest emissions and are used primarily to identify elements up to atomic number 30; the L and M lines are used for elements greater than atomic number 30.

A typical accelerating voltage for good X-ray analysis is 20 KV. Other important factors which can affect the quality of the EDX analysis include count rate (deadtime), specimen topography, detector geometry, and the specimen coating. For additional information on how these factors affect the EDX analysis, see Postek et al (1980), chapters 4 and 5. This atlas deals only with elemental analysis of rough-cut samples; thus the data obtained is at best only semiquantitative.

During EDX analysis the mineral to be identified is isolated in the SEM at approximately 20,000× to 50,000× (or in a reduced area mode). All X-rays generated from the isolated area are collected and separated by energy level in a multichannel pulse height analyzer. Any major element in the sample (above Z = 11, Na, sodium) will yield a peak on a graph (the EDX spectrum) at its unique energy level. The majority of peaks found in rock-forming minerals will be common elements such as silicon (Si), aluminum (Al), magnesium (Mg), iron (Fe), sodium (Na), potassium (K), calcium (Ca), titanium (Ti), and sulfur (S), and no elements below Na (Z = 11) will be detected. Peaks representing gold (Au), copper (Cu), and palladium (Pd)

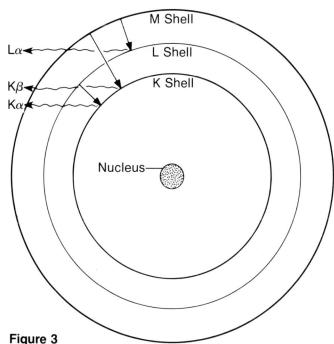

Figure 3

Electron Transitions in an Atom (modified from Goldstein and Yakowitz, 1978). When an orbiting electron is ejected from the K shell by the SEM electron beam, to regain stability an electron from the L shell fills the vacancy. The amount of x-ray energy released during this transition is termed the Kα x-ray. If an electron from the M shell fills the vacancy, the energy released is termed Kβ etc.

usually indicate radiation from the metal coatings and the specimen plug, so should be ignored. Only elemental concentrations above approximately 1% are displayed.

Identification of each peak on the EDX spectrum is done by lining up the apex with the energy scale (KeV) at the bottom of the graph (Figure 4). This number is then compared with the X-ray energy tables (see back of this atlas), which list the major X-ray energies for each element. Today all commercial EDX systems are equipped with a preprogrammed MKL marker system to aid in rapid identification of the displayed peaks. By simply dialing in the atomic number of any element, a cursor will appear on the screen indicating the major peak positions for that element.

After all peaks on the EDX spectrum are identified, the relative concentrations of the elements are then compared with the crystal morphology and the chemical formula of the suspected mineral. Correlation of the peak heights of Si, Al, K, and Ca with the chemical formula is possible because the peak heights are roughly proportional to their concentration. However,

Figure 4 — How To Identify Peaks in the EDX Spectrum

Potassium Feldspar KAl Si$_3$ O$_8$

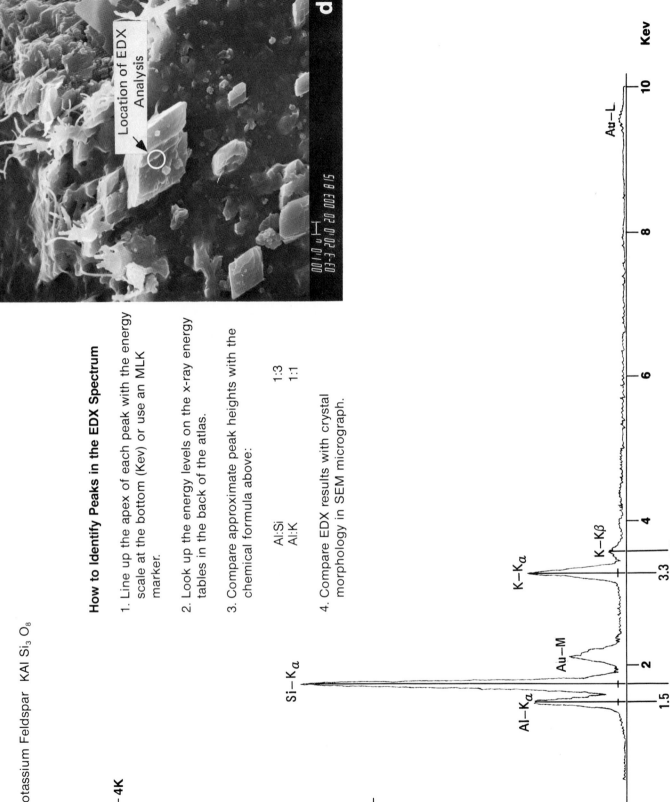

How to Identify Peaks in the EDX Spectrum

1. Line up the apex of each peak with the energy scale at the bottom (Kev) or use an MLK marker.

2. Look up the energy levels on the x-ray energy tables in the back of the atlas.

3. Compare approximate peak heights with the chemical formula above:

   Al:Si   1:3
   Al:K   1:1

4. Compare EDX results with crystal morphology in SEM micrograph.

problems with comparing peak heights to concentration do occur at both the low- and high-atomic number ends. For example, Na and Mg peak heights are always reduced relative to their concentration due to absorption of these low-energy X-rays in the Beryllium window of the detector. However, the establishment of at least the presence of these elements aids in the mineral identification.

Two other potential problems in EDX interpretation occur due to poor resolution and electron beam penetration through very thin materials, such as clays. Most EDX systems have resolutions better than 150 eV (measured at the 5.89 KeV Mn K_α energy level). If the X-ray energies of two elements in a mineral are less than 150 eV apart, this will appear on the EDX spectrum as a single asymmetrical peak, rather than two discrete peaks at that energy level. For example, the mineral Florencite contains phosphorus (P) and is coated with gold (Au) for SEM analysis (see EDX spectrum of Florencite in this atlas). Since both P and Au have major emission lines around 2.1 to 2.2 KeV, the Florencite spectrum shows only one peak separated at the apex into two points, one representing P and the other Au. Thus, if any peak in the spectrum does not have the ideal gaussian shape, suspect the presence of more than one element.

Another problem occurs when trying to identify very thin minerals (e.g. authigenic illite). The electron beam is strong enough to penetrate through a thin mineral into any underlying detrital grain. This results in the weak detection of elements from the underlying grain in addition to the elements from the thin mineral you are trying to identify. The EDX spectrum obtained then represents a composite of both minerals. Therefore, to correctly identify thin minerals, probe the thickest areas, identify the underlying grain and mentally subtract the possible elemental contribution of the underlying grain from the EDX spectrum.

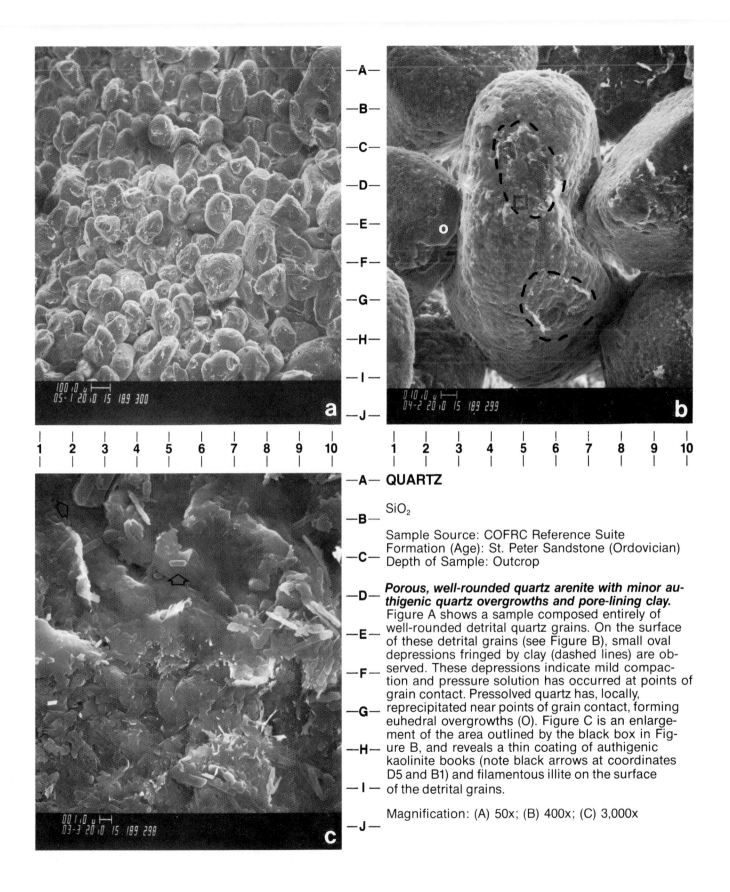

-A- QUARTZ

SiO₂

Sample Source: COFRC Reference Suite
Formation (Age): St. Peter Sandstone (Ordovician)
Depth of Sample: Outcrop

Porous, well-rounded quartz arenite with minor authigenic quartz overgrowths and pore-lining clay.
Figure A shows a sample composed entirely of well-rounded detrital quartz grains. On the surface of these detrital grains (see Figure B), small oval depressions fringed by clay (dashed lines) are observed. These depressions indicate mild compaction and pressure solution has occurred at points of grain contact. Pressolved quartz has, locally, reprecipitated near points of grain contact, forming euhedral overgrowths (O). Figure C is an enlargement of the area outlined by the black box in Figure B, and reveals a thin coating of authigenic kaolinite books (note black arrows at coordinates D5 and B1) and filamentous illite on the surface of the detrital grains.

Magnification: (A) 50x; (B) 400x; (C) 3,000x

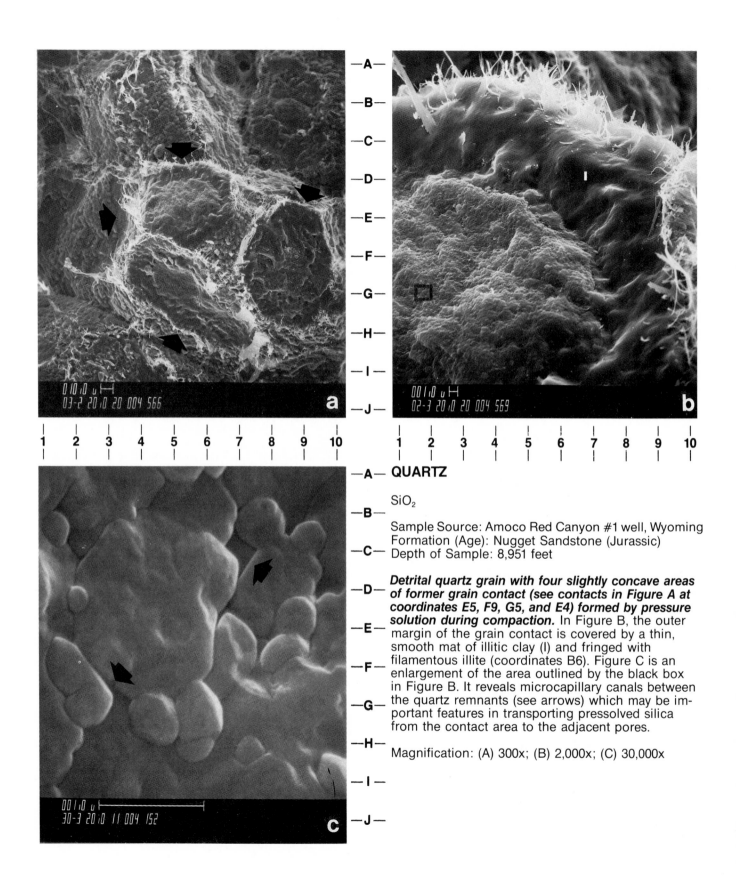

QUARTZ

SiO$_2$

Sample Source: Amoco Red Canyon #1 well, Wyoming
Formation (Age): Nugget Sandstone (Jurassic)
Depth of Sample: 8,951 feet

***Detrital quartz grain with four slightly concave areas
of former grain contact (see contacts in Figure A at
coordinates E5, F9, G5, and E4) formed by pressure
solution during compaction.*** In Figure B, the outer
margin of the grain contact is covered by a thin,
smooth mat of illitic clay (I) and fringed with
filamentous illite (coordinates B6). Figure C is an
enlargement of the area outlined by the black box
in Figure B. It reveals microcapillary canals between
the quartz remnants (see arrows) which may be important features in transporting pressolved silica
from the contact area to the adjacent pores.

Magnification: (A) 300x; (B) 2,000x; (C) 30,000x

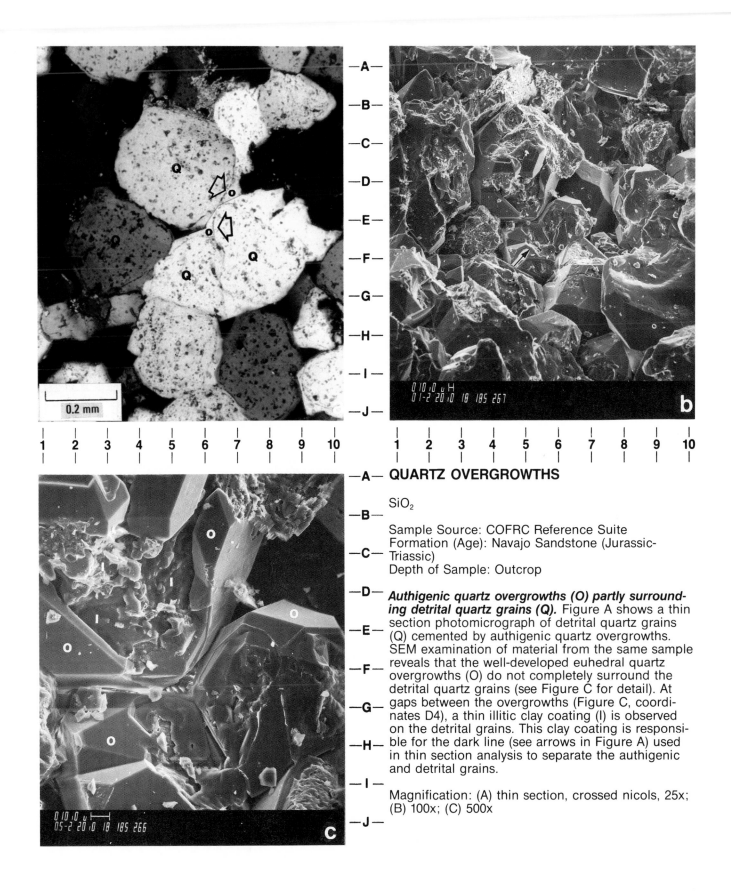

QUARTZ OVERGROWTHS

SiO$_2$

Sample Source: COFRC Reference Suite
Formation (Age): Navajo Sandstone (Jurassic-Triassic)
Depth of Sample: Outcrop

Authigenic quartz overgrowths (O) partly surrounding detrital quartz grains (Q). Figure A shows a thin section photomicrograph of detrital quartz grains (Q) cemented by authigenic quartz overgrowths. SEM examination of material from the same sample reveals that the well-developed euhedral quartz overgrowths (O) do not completely surround the detrital quartz grains (see Figure C for detail). At gaps between the overgrowths (Figure C, coordinates D4), a thin illitic clay coating (I) is observed on the detrital grains. This clay coating is responsible for the dark line (see arrows in Figure A) used in thin section analysis to separate the authigenic and detrital grains.

Magnification: (A) thin section, crossed nicols, 25x; (B) 100x; (C) 500x

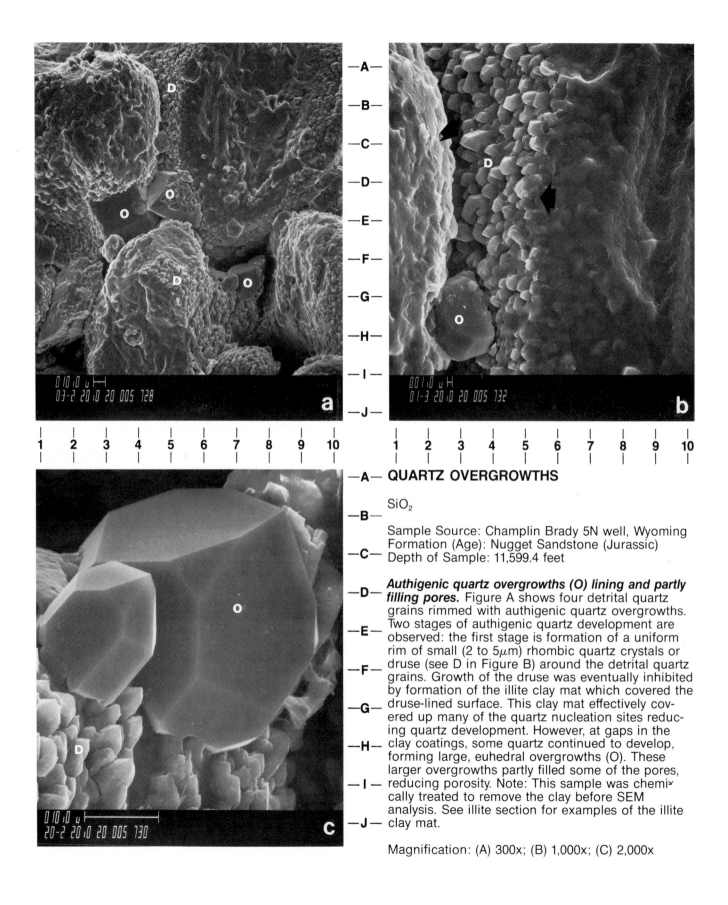

QUARTZ OVERGROWTHS

SiO$_2$

Sample Source: Champlin Brady 5N well, Wyoming
Formation (Age): Nugget Sandstone (Jurassic)
Depth of Sample: 11,599.4 feet

Authigenic quartz overgrowths (O) lining and partly filling pores. Figure A shows four detrital quartz grains rimmed with authigenic quartz overgrowths. Two stages of authigenic quartz development are observed: the first stage is formation of a uniform rim of small (2 to 5μm) rhombic quartz crystals or druse (see D in Figure B) around the detrital quartz grains. Growth of the druse was eventually inhibited by formation of the illite clay mat which covered the druse-lined surface. This clay mat effectively covered up many of the quartz nucleation sites reducing quartz development. However, at gaps in the clay coatings, some quartz continued to develop, forming large, euhedral overgrowths (O). These larger overgrowths partly filled some of the pores, reducing porosity. Note: This sample was chemically treated to remove the clay before SEM analysis. See illite section for examples of the illite clay mat.

Magnification: (A) 300x; (B) 1,000x; (C) 2,000x

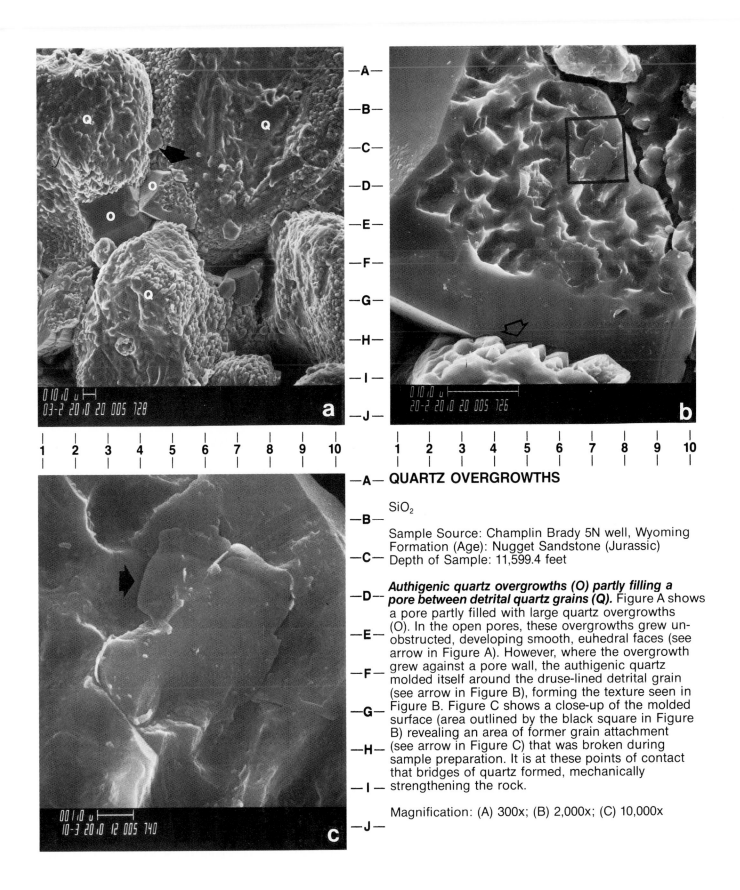

QUARTZ OVERGROWTHS

SiO_2

Sample Source: Champlin Brady 5N well, Wyoming
Formation (Age): Nugget Sandstone (Jurassic)
Depth of Sample: 11,599.4 feet

Authigenic quartz overgrowths (O) partly filling a pore between detrital quartz grains (Q). Figure A shows a pore partly filled with large quartz overgrowths (O). In the open pores, these overgrowths grew unobstructed, developing smooth, euhedral faces (see arrow in Figure A). However, where the overgrowth grew against a pore wall, the authigenic quartz molded itself around the druse-lined detrital grain (see arrow in Figure B), forming the texture seen in Figure B. Figure C shows a close-up of the molded surface (area outlined by the black square in Figure B) revealing an area of former grain attachment (see arrow in Figure C) that was broken during sample preparation. It is at these points of contact that bridges of quartz formed, mechanically strengthening the rock.

Magnification: (A) 300x; (B) 2,000x; (C) 10,000x

CRISTOBALITE (OPAL-CT)

SiO$_2$

Sample Source: Pismo Beach, California
Formation (Age): Monterey Formation (Miocene)
Depth of Sample: Outcrop

Lepispheres composed of bladed, authigenic crystals of cristobalite (C) lining cavities in a porcellanite. The bladed morphology of the lepispheres and X-ray diffraction analysis combined with an EDX spectrum yielding only silica (see EDX on facing page) were used to identify this mineral as cristobalite. In this example, the cristobalite is associated with elongate, fibrous rods of the zeolite, mordenite (see M in Figure A; arrow in Figure B at coordinates C7.5; arrows in Figure C at coordinates F4.5 and E8) and clusters of siderite (S).

Magnification: (A) 1,000x; (B) 5,000x; (C) 10,000x; (D) 20,000x

Energy Dispersive X—Ray Spectrum (EDX)

Cristobalite (Opal-CT) SiO$_2$

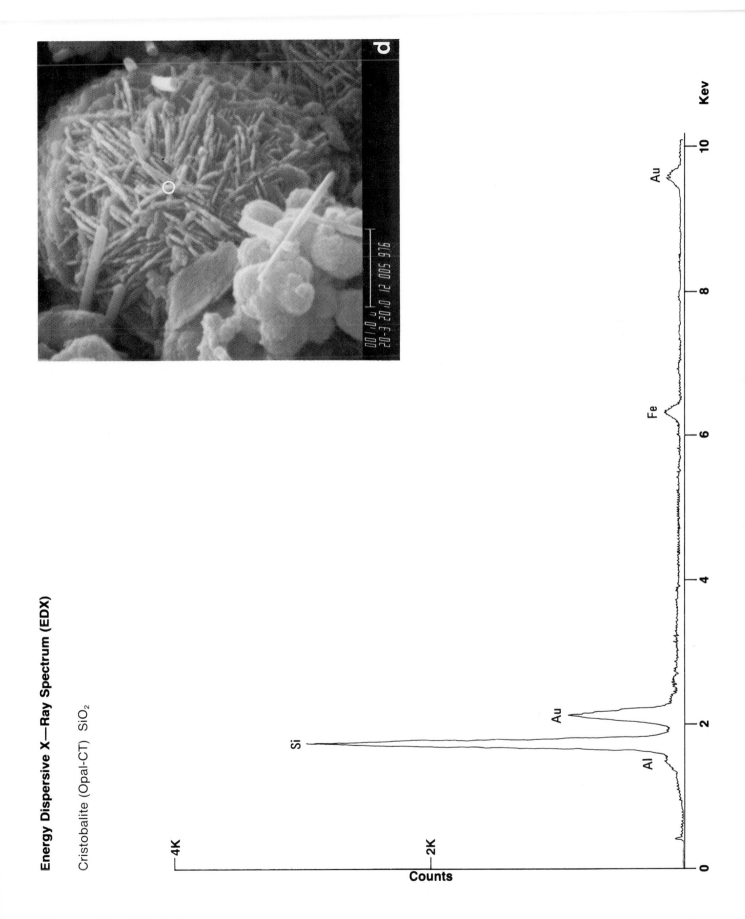

Silicates—Silica (Cristobalite)

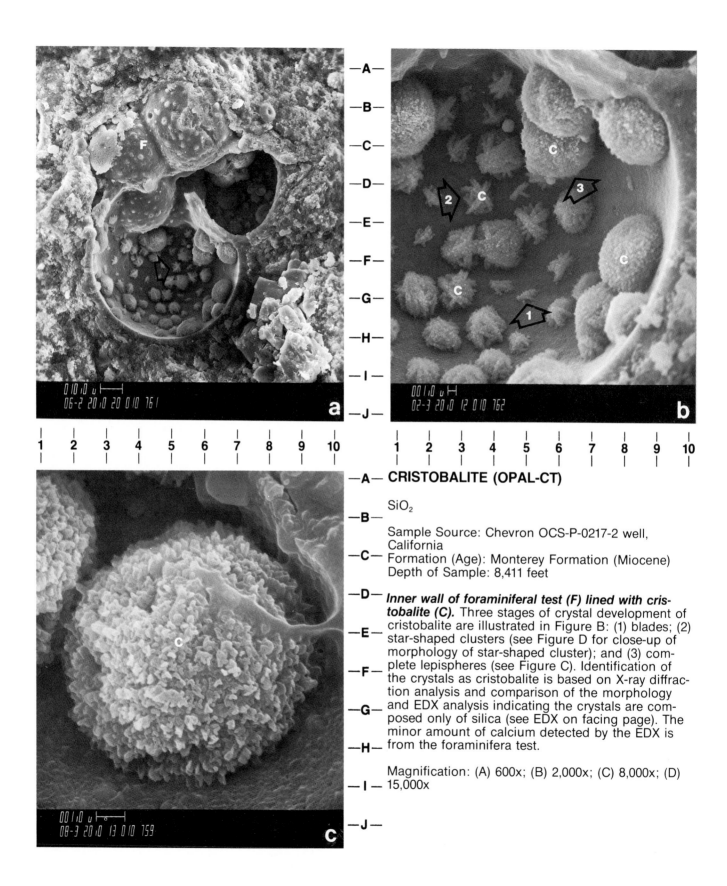

CRISTOBALITE (OPAL-CT)

SiO$_2$

Sample Source: Chevron OCS-P-0217-2 well, California
Formation (Age): Monterey Formation (Miocene)
Depth of Sample: 8,411 feet

Inner wall of foraminiferal test (F) lined with cristobalite (C). Three stages of crystal development of cristobalite are illustrated in Figure B: (1) blades; (2) star-shaped clusters (see Figure D for close-up of morphology of star-shaped cluster); and (3) complete lepispheres (see Figure C). Identification of the crystals as cristobalite is based on X-ray diffraction analysis and comparison of the morphology and EDX analysis indicating the crystals are composed only of silica (see EDX on facing page). The minor amount of calcium detected by the EDX is from the foraminifera test.

Magnification: (A) 600x; (B) 2,000x; (C) 8,000x; (D) 15,000x

Energy Dispersive X—Ray Spectrum (EDX)

Cristobalite (Opal-CT) Si O$_2$

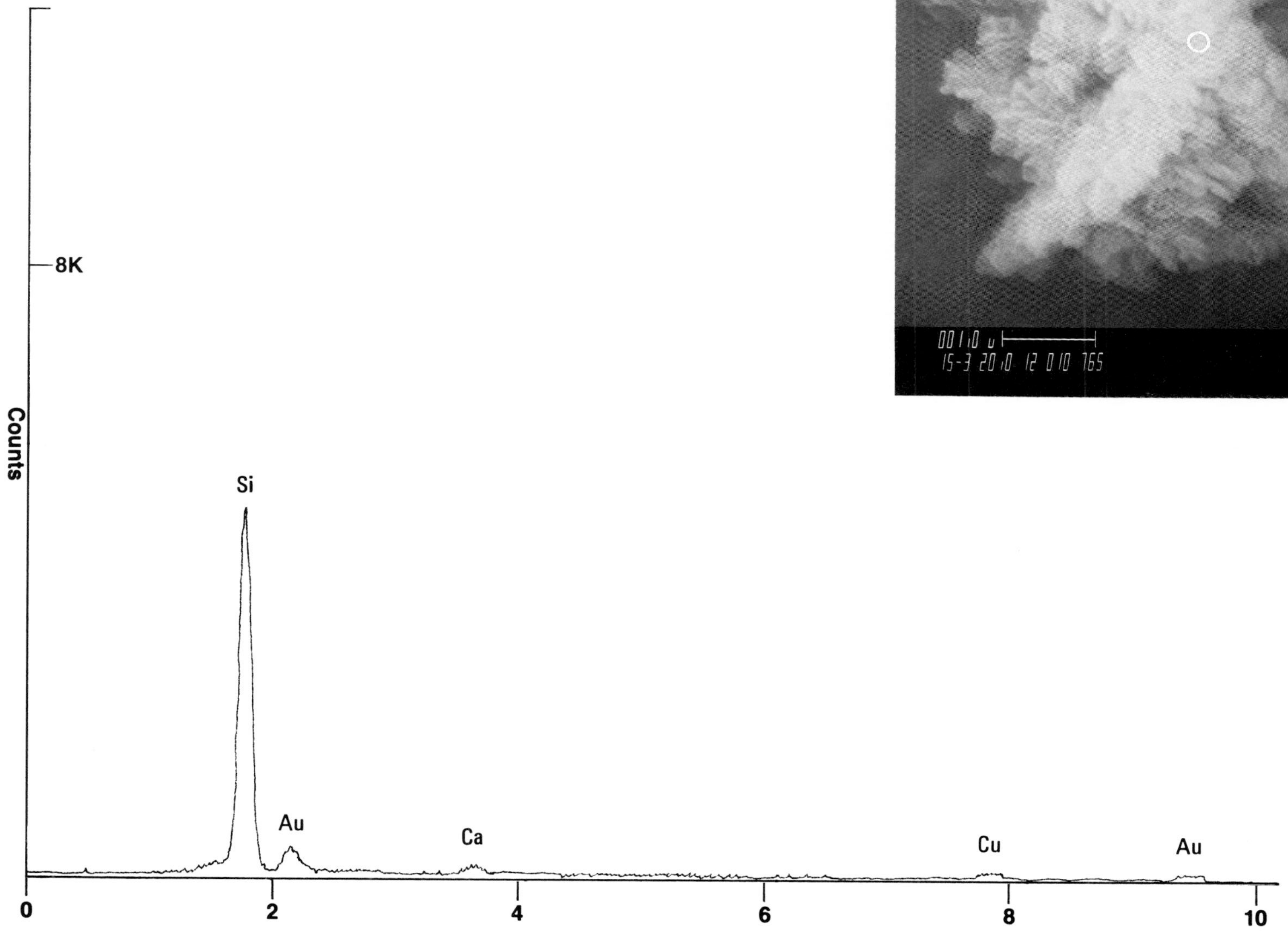

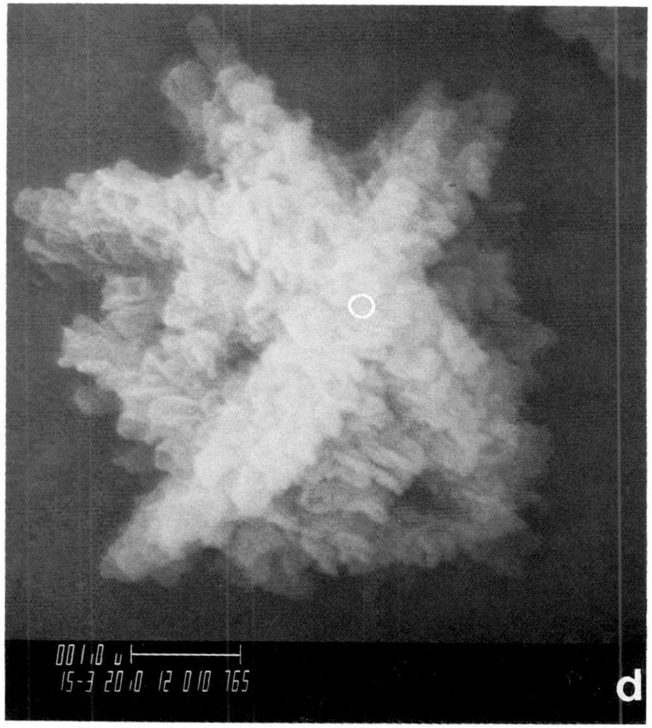

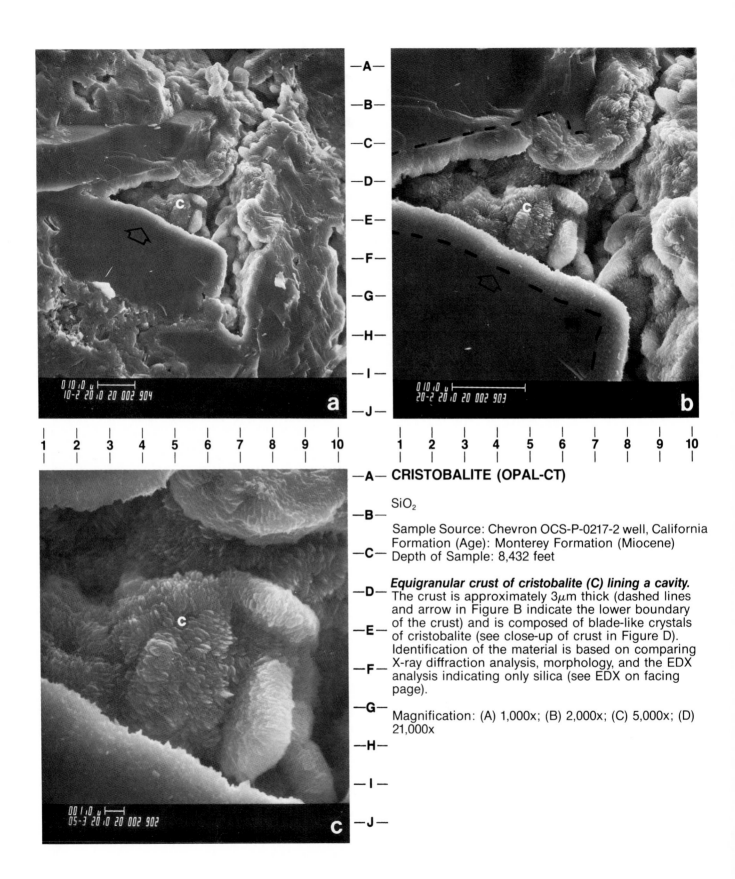

CRISTOBALITE (OPAL-CT)

SiO_2

Sample Source: Chevron OCS-P-0217-2 well, California
Formation (Age): Monterey Formation (Miocene)
Depth of Sample: 8,432 feet

Equigranular crust of cristobalite (C) lining a cavity.
The crust is approximately $3\mu m$ thick (dashed lines
and arrow in Figure B indicate the lower boundary
of the crust) and is composed of blade-like crystals
of cristobalite (see close-up of crust in Figure D).
Identification of the material is based on comparing
X-ray diffraction analysis, morphology, and the EDX
analysis indicating only silica (see EDX on facing
page).

Magnification: (A) 1,000x; (B) 2,000x; (C) 5,000x; (D)
21,000x

Energy Dispersive X—Ray Spectrum (EDX)

Cristobalite (Opal-CT) SiO_2

Silicates—Silica (Cristobalite)

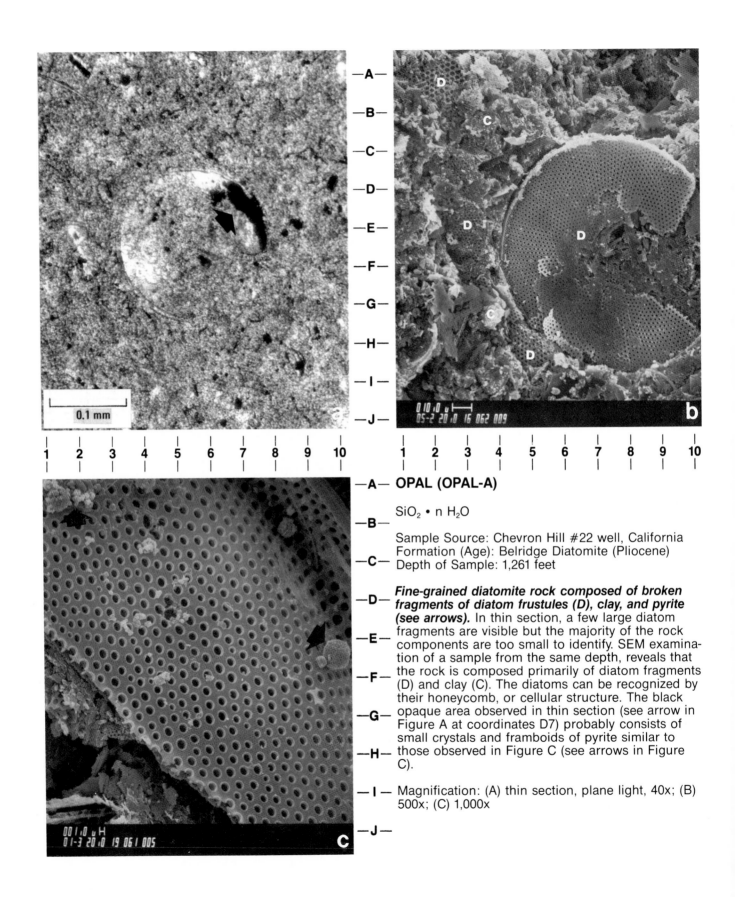

—A—

—B—

—C—

—D—

—E—

—F—

—G—

—H—

—I—

—J—

—A— **OPAL (OPAL-A)**

—B— $SiO_2 \cdot n\ H_2O$

—C— Sample Source: Chevron Hill #22 well, California
Formation (Age): Belridge Diatomite (Pliocene)
Depth of Sample: 1,261 feet

—D— ***Fine-grained diatomite rock composed of broken
fragments of diatom frustules (D), clay, and pyrite
(see arrows).*** In thin section, a few large diatom
fragments are visible but the majority of the rock
components are too small to identify. SEM examination of a sample from the same depth, reveals that
—E— the rock is composed primarily of diatom fragments
(D) and clay (C). The diatoms can be recognized by
their honeycomb, or cellular structure. The black
—F— opaque area observed in thin section (see arrow in
Figure A at coordinates D7) probably consists of
small crystals and framboids of pyrite similar to
—G— those observed in Figure C (see arrows in Figure
C).

—H—

—I— Magnification: (A) thin section, plane light, 40x; (B)
500x; (C) 1,000x

—J—

Silicates—Silica (Opal)

Feldspars

—A— **POTASSIUM FELDSPAR**

KAlSi$_3$O$_8$

Sample Source: Amoco Champlin #224A-1 well, Wyoming
Formation (Age): Nugget Sandstone (Jurassic)
Depth of Sample: 7,550 feet

Authigenic K-feldspar overgrowths on a detrital K-feldspar grain. Figure A shows a thin section of a well-rounded detrital K-feldspar (microcline) grain rimmed with small, jagged overgrowths (see arrow on Figure A at coordinates D4.5). In the SEM, these overgrowths appear as small (2 to 10μm) rhombic crystals (see arrows in Figure C at coordinates C2, C6, and F4.5) partly covering a detrital K-feldspar surface (see Figure D for a close-up of one of these overgrowths). Bald areas, devoid of overgrowths, are areas of former grain contact (outlined by dashed lines in Figure C). Identification of both the detrital and authigenic minerals is based on EDX analysis yielding the major elements Si, Al, and K (see EDX on facing page). This is a typical EDX spectrum for K-feldspar.

Magnification: (A) thin section, crossed nicols, 64x; (B) 100x; (C) 400x; (D) 3,000x

Silicates—Feldspar (Potassium)

Energy Dispersive X—Ray Spectrum (EDX)

Potassium Feldspar K Al Si$_3$ O$_8$

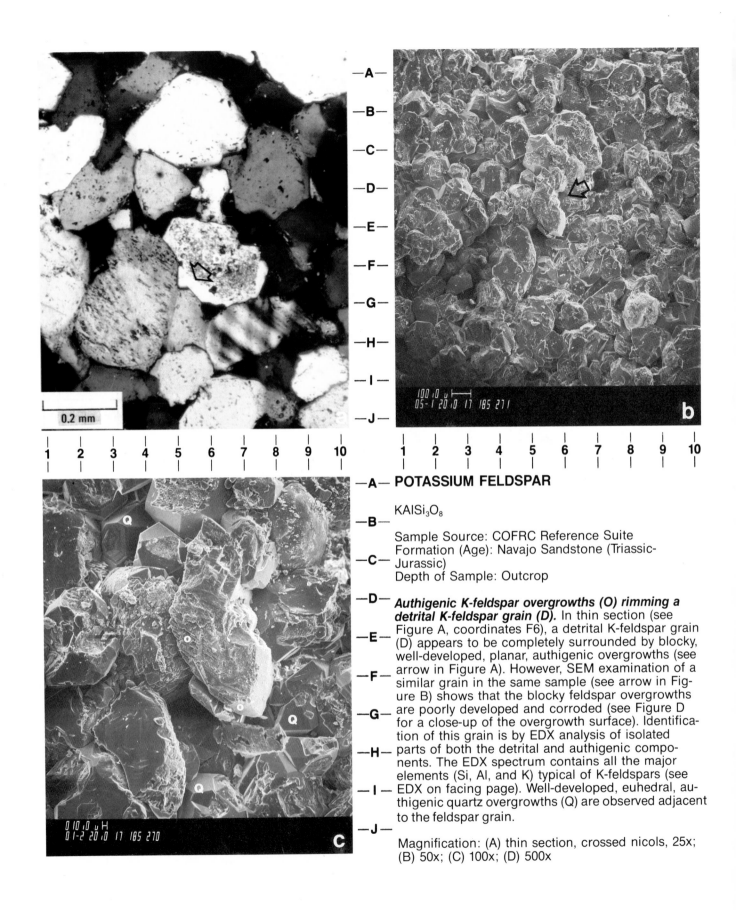

POTASSIUM FELDSPAR

KAlSi$_3$O$_8$

Sample Source: COFRC Reference Suite
Formation (Age): Navajo Sandstone (Triassic-Jurassic)
Depth of Sample: Outcrop

Authigenic K-feldspar overgrowths (O) rimming a detrital K-feldspar grain (D). In thin section (see Figure A, coordinates F6), a detrital K-feldspar grain (D) appears to be completely surrounded by blocky, well-developed, planar, authigenic overgrowths (see arrow in Figure A). However, SEM examination of a similar grain in the same sample (see arrow in Figure B) shows that the blocky feldspar overgrowths are poorly developed and corroded (see Figure D for a close-up of the overgrowth surface). Identification of this grain is by EDX analysis of isolated parts of both the detrital and authigenic components. The EDX spectrum contains all the major elements (Si, Al, and K) typical of K-feldspars (see EDX on facing page). Well-developed, euhedral, authigenic quartz overgrowths (Q) are observed adjacent to the feldspar grain.

Magnification: (A) thin section, crossed nicols, 25x; (B) 50x; (C) 100x; (D) 500x

Silicates—Feldspar (Potassium)

Energy Dispersive X—Ray Spectrum (EDX)

Potassium Feldspar K Al Si$_3$ O$_8$

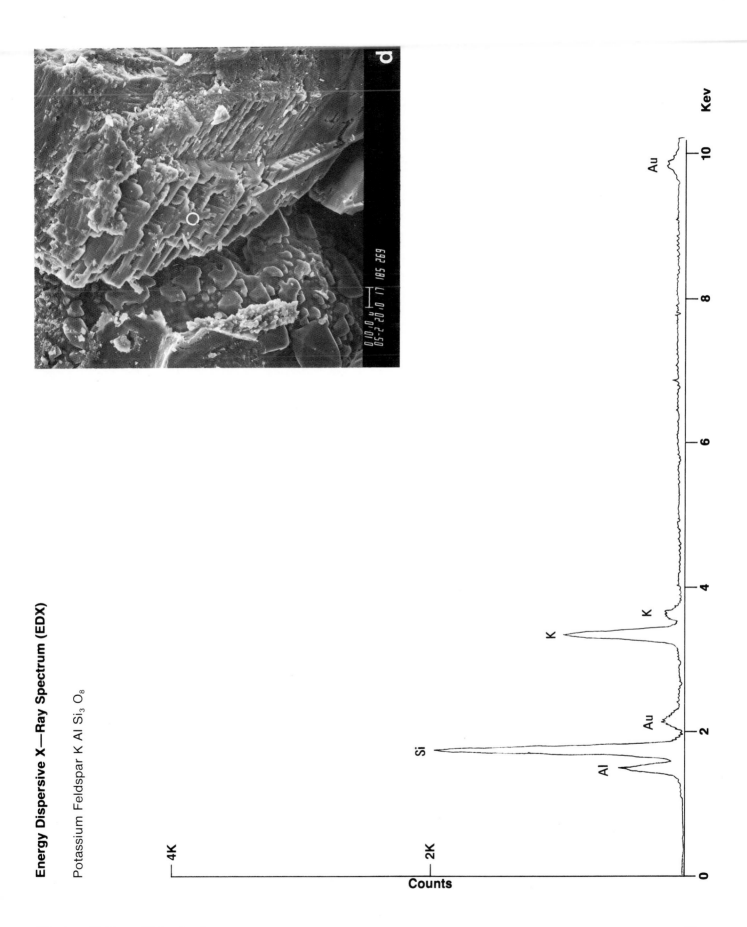

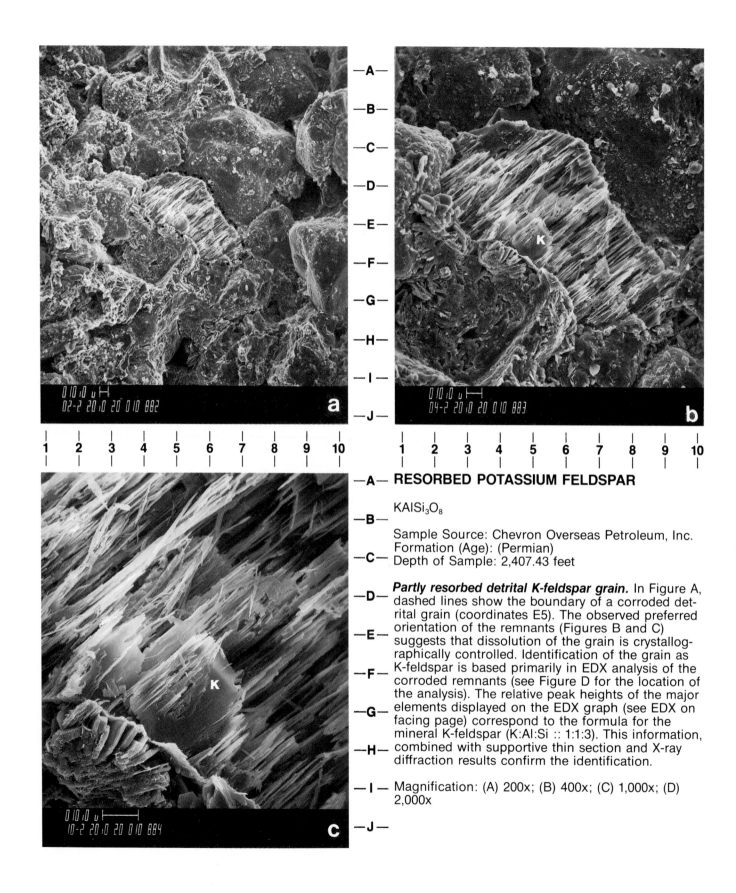

RESORBED POTASSIUM FELDSPAR

$KAlSi_3O_8$

Sample Source: Chevron Overseas Petroleum, Inc.
Formation (Age): (Permian)
Depth of Sample: 2,407.43 feet

Partly resorbed detrital K-feldspar grain. In Figure A,
dashed lines show the boundary of a corroded detrital grain (coordinates E5). The observed preferred
orientation of the remnants (Figures B and C)
suggests that dissolution of the grain is crystallog-
raphically controlled. Identification of the grain as
K-feldspar is based primarily in EDX analysis of the
corroded remnants (see Figure D for the location of
the analysis). The relative peak heights of the major
elements displayed on the EDX graph (see EDX on
facing page) correspond to the formula for the
mineral K-feldspar (K:Al:Si :: 1:1:3). This information,
combined with supportive thin section and X-ray
diffraction results confirm the identification.

Magnification: (A) 200x; (B) 400x; (C) 1,000x; (D)
2,000x

Energy Dispersive X—Ray Spectrum (EDX)

Potassium Feldspar K Al Si$_3$ O$_8$

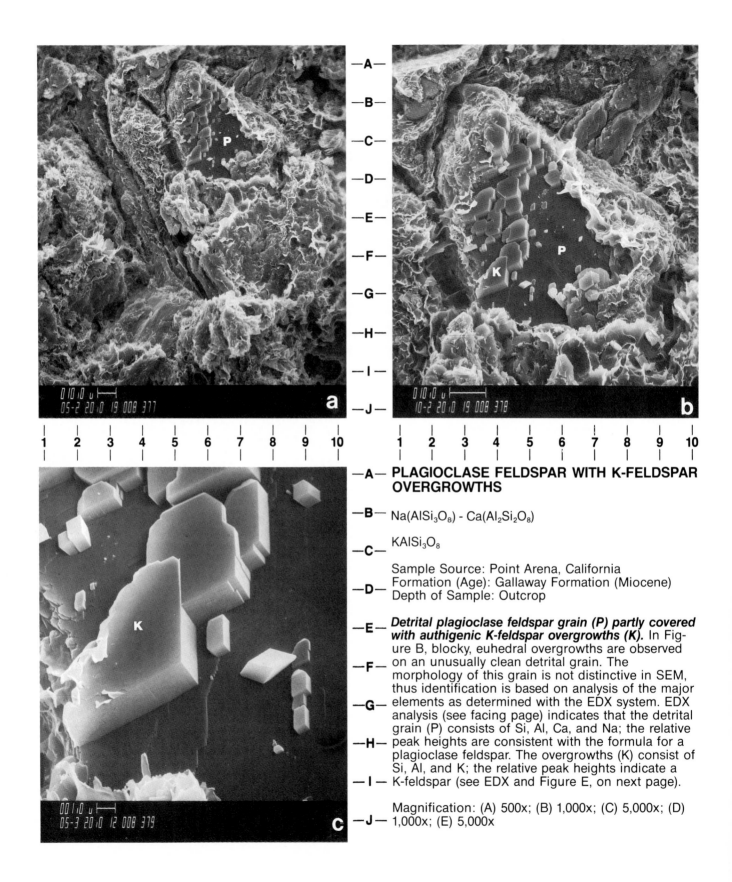

1 2 3 4 5 6 7 8 9 10

A— PLAGIOCLASE FELDSPAR WITH K-FELDSPAR OVERGROWTHS

B— Na(AlSi$_3$O$_8$) - Ca(Al$_2$Si$_2$O$_8$)

C— KAlSi$_3$O$_8$

D— Sample Source: Point Arena, California
Formation (Age): Gallaway Formation (Miocene)
Depth of Sample: Outcrop

E— *Detrital plagioclase feldspar grain (P) partly covered with authigenic K-feldspar overgrowths (K).* In Figure B, blocky, euhedral overgrowths are observed on an unusually clean detrital grain. The morphology of this grain is not distinctive in SEM, thus identification is based on analysis of the major elements as determined with the EDX system. EDX analysis (see facing page) indicates that the detrital grain (P) consists of Si, Al, Ca, and Na; the relative peak heights are consistent with the formula for a plagioclase feldspar. The overgrowths (K) consist of Si, Al, and K; the relative peak heights indicate a K-feldspar (see EDX and Figure E, on next page).

Magnification: (A) 500x; (B) 1,000x; (C) 5,000x; (D) 1,000x; (E) 5,000x

Energy Dispersive X—Ray Spectrum (EDX)

Plagioclase Feldspar Na (Al Si$_3$ O$_8$) — Ca (Al$_2$ Si$_2$ O$_8$)

Silicates—Feldspar (Plagioclase)

Energy Dispersive X—Ray Spectrum (EDX)

Potassium Feldspar K Al Si$_3$ O$_8$

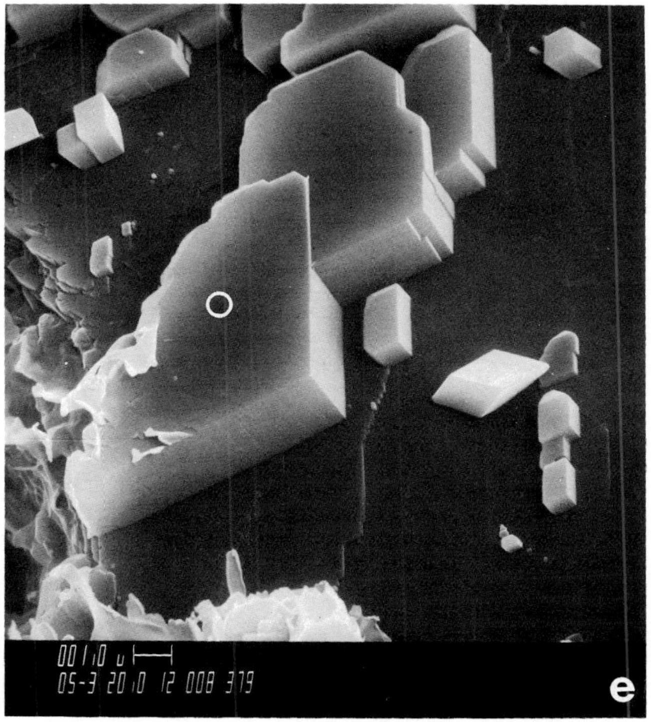

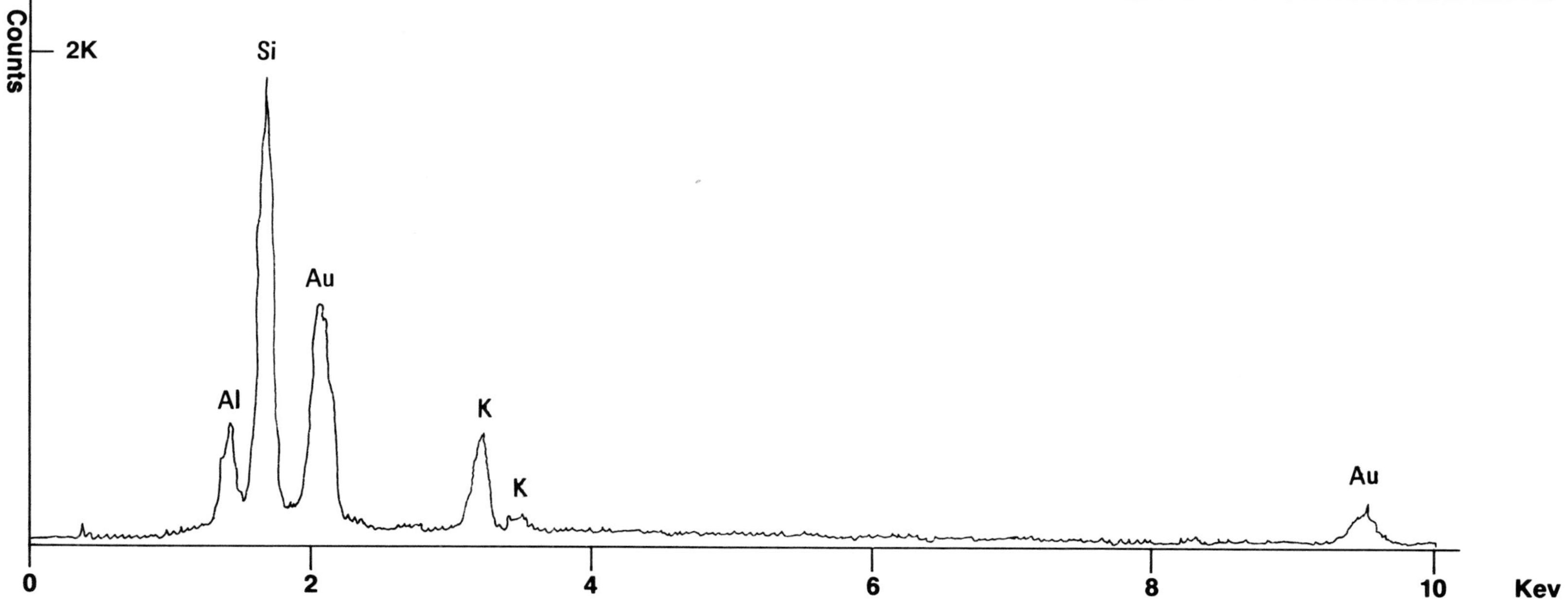

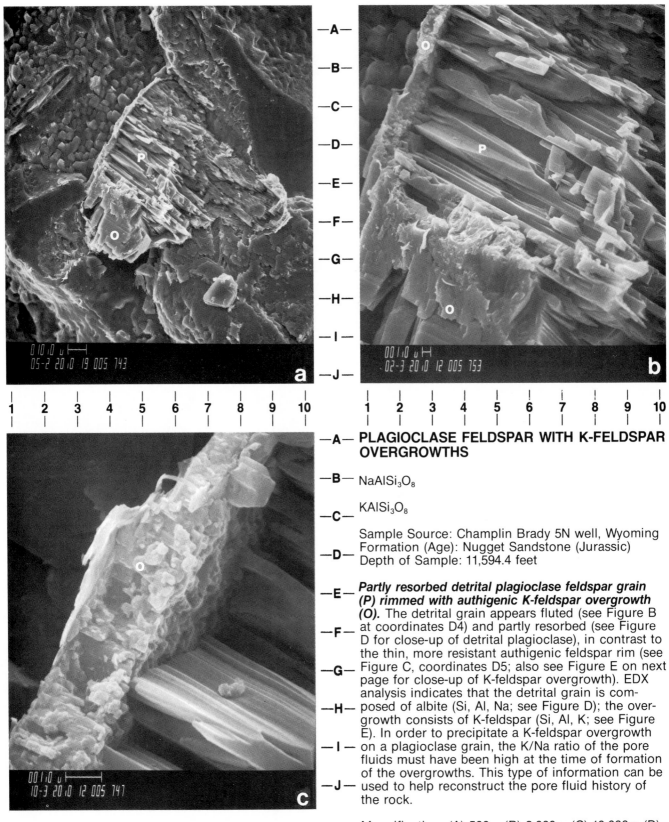

PLAGIOCLASE FELDSPAR WITH K-FELDSPAR OVERGROWTHS

NaAlSi$_3$O$_8$

KAlSi$_3$O$_8$

Sample Source: Champlin Brady 5N well, Wyoming
Formation (Age): Nugget Sandstone (Jurassic)
Depth of Sample: 11,594.4 feet

Partly resorbed detrital plagioclase feldspar grain (P) rimmed with authigenic K-feldspar overgrowth (O). The detrital grain appears fluted (see Figure B at coordinates D4) and partly resorbed (see Figure D for close-up of detrital plagioclase), in contrast to the thin, more resistant authigenic feldspar rim (see Figure C, coordinates D5; also see Figure E on next page for close-up of K-feldspar overgrowth). EDX analysis indicates that the detrital grain is composed of albite (Si, Al, Na; see Figure D); the overgrowth consists of K-feldspar (Si, Al, K; see Figure E). In order to precipitate a K-feldspar overgrowth on a plagioclase grain, the K/Na ratio of the pore fluids must have been high at the time of formation of the overgrowths. This type of information can be used to help reconstruct the pore fluid history of the rock.

Magnification: (A) 500x; (B) 2,000x; (C) 10,000x; (D) 10,000x; (E) 20,000x

Silicates—Feldspar (Plagioclase)

Energy Dispersive X—Ray Spectrum (EDX)

Albite Na (Al Si$_3$ O$_8$)

Energy Dispersive X—Ray Spectrum (EDX)

Potassium Feldspar K Al Si$_3$ O$_8$

Silicates—Feldspar (Plagioclase)

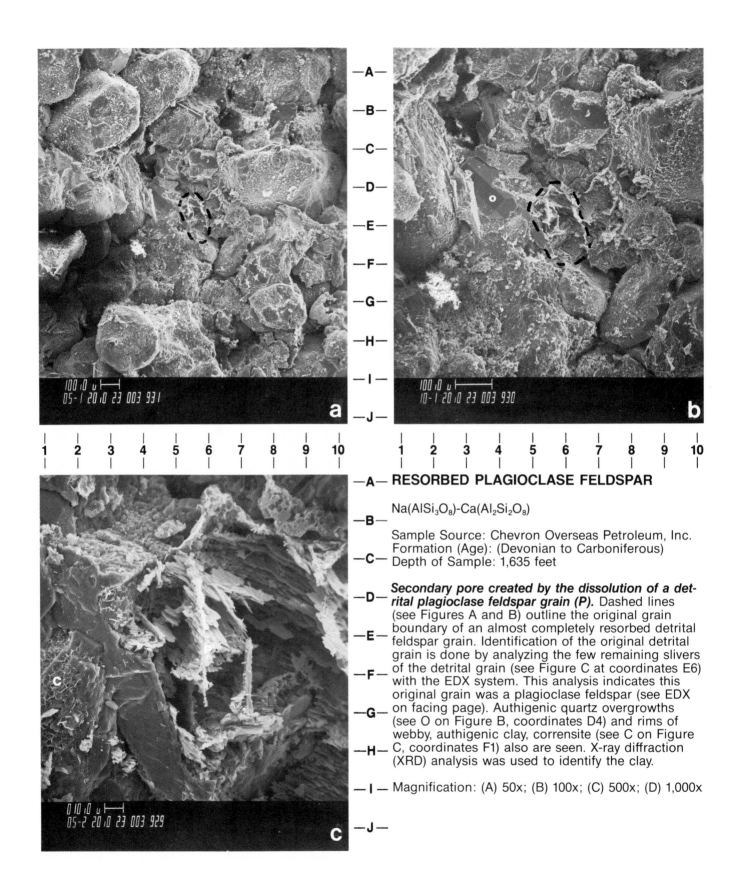

Grid labels (right of figures a and b): —A— —B— —C— —D— —E— —F— —G— —H— —I— —J—

Scale labels under figures: 1 2 3 4 5 6 7 8 9 10

Figure a label (in image): 100,0 u 05-1 20,0 23 003 931 — **a**

Figure b label (in image): 100,0 u 10-1 20,0 23 003 930 — **b**

Figure c label (in image): 0 10,0 u 05-2 20,0 23 003 929 — **c**

—A— **RESORBED PLAGIOCLASE FELDSPAR**

—B— $Na(AlSi_3O_8)-Ca(Al_2Si_2O_8)$

Sample Source: Chevron Overseas Petroleum, Inc.
Formation (Age): (Devonian to Carboniferous)
—C— Depth of Sample: 1,635 feet

—D— ***Secondary pore created by the dissolution of a detrital plagioclase feldspar grain (P).*** Dashed lines (see Figures A and B) outline the original grain boundary of an almost completely resorbed detrital —E— feldspar grain. Identification of the original detrital grain is done by analyzing the few remaining slivers of the detrital grain (see Figure C at coordinates E6) —F— with the EDX system. This analysis indicates this original grain was a plagioclase feldspar (see EDX on facing page). Authigenic quartz overgrowths —G— (see O on Figure B, coordinates D4) and rims of webby, authigenic clay, corrensite (see C on Figure C, coordinates F1) also are seen. X-ray diffraction —H— (XRD) analysis was used to identify the clay.

—I— Magnification: (A) 50x; (B) 100x; (C) 500x; (D) 1,000x

—J—

Silicates—Feldspar (Plagioclase)

Energy Dispersive X—Ray Spectrum (EDX)

Plagioclase Feldspar Na (Al Si$_3$ O$_8$) — Ca (Al$_2$ Si$_2$ O$_8$)

Silicates—Feldspar (Plagioclase)

Clays

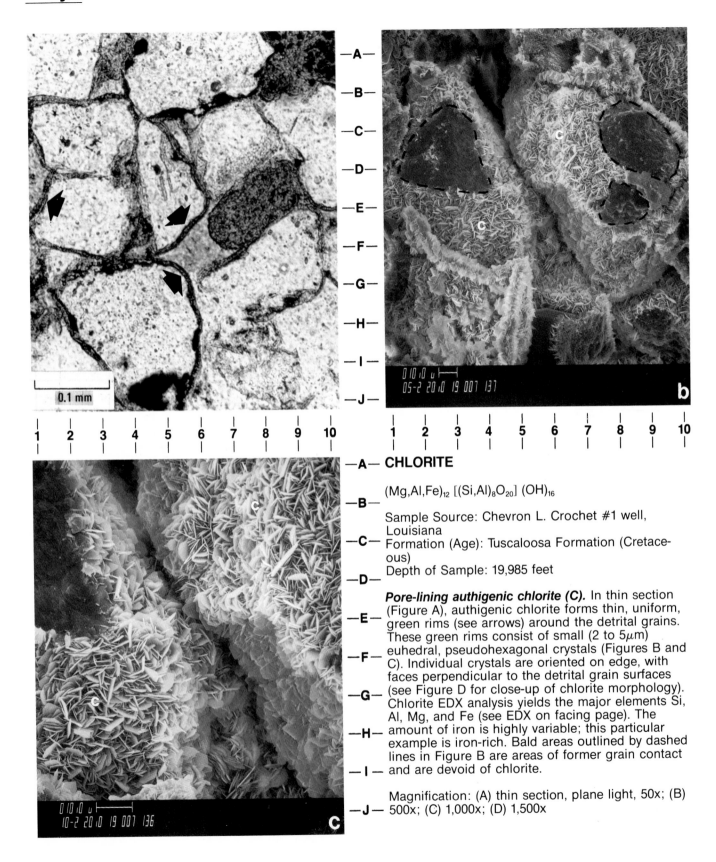

—A— **CHLORITE**

—B— $(Mg,Al,Fe)_{12} [(Si,Al)_8 O_{20}] (OH)_{16}$

Sample Source: Chevron L. Crochet #1 well, Louisiana

—C— Formation (Age): Tuscaloosa Formation (Cretaceous)

Depth of Sample: 19,985 feet

—D—

Pore-lining authigenic chlorite (C). In thin section (Figure A), authigenic chlorite forms thin, uniform, green rims (see arrows) around the detrital grains. These green rims consist of small (2 to 5μm) euhedral, pseudohexagonal crystals (Figures B and C). Individual crystals are oriented on edge, with faces perpendicular to the detrital grain surfaces (see Figure D for close-up of chlorite morphology). Chlorite EDX analysis yields the major elements Si, Al, Mg, and Fe (see EDX on facing page). The amount of iron is highly variable; this particular example is iron-rich. Bald areas outlined by dashed lines in Figure B are areas of former grain contact and are devoid of chlorite.

Magnification: (A) thin section, plane light, 50x; (B) 500x; (C) 1,000x; (D) 1,500x

Energy Dispersive X—Ray Spectrum (EDX)

Chlorite $(Mg, Al, Fe)_{12}$ $[(Si, Al)_8$ $O_{20}]$ $(OH)_{16}$

Silicates—Clay (Chlorite)

CHLORITE

$(Mg,Al,Fe)_{12} [(Si,Al)_8O_{20}] (OH)_{16}$

Sample Source: Chevron L. Crochet #1 well, Louisiana
Formation (Age): Tuscaloosa Formation (Cretaceous)
Depth of Sample: 19,985 feet

Authigenic, pore-lining chlorite rosettes. In Figure A, individual chlorite platelets (arrow) are arranged in a rosette pattern on a chlorite-coated detrital quartz grain. These euhedral, pseudohexagonal crystals are approximately 2 to 5μm in diameter and less than 1μm thick (see Figures B and C). EDX analysis indicates an iron-rich chlorite containing the major elements Si, Al, Mg, and Fe (see EDX on facing page). In this case, EDX analysis of the major elements and their relative peak heights are very important; otherwise this mineral might be confused with other rosette-like minerals.

Magnification: (A) 1,000x; (B) 6,000x; (C) 10,000x; (D) 20,000x

Energy Dispersive X—Ray Spectrum (EDX)

Chlorite $(Mg, Al, Fe)_{12}[(Si, Al)_8 \ O_{20}] \ (OH)_{16}$

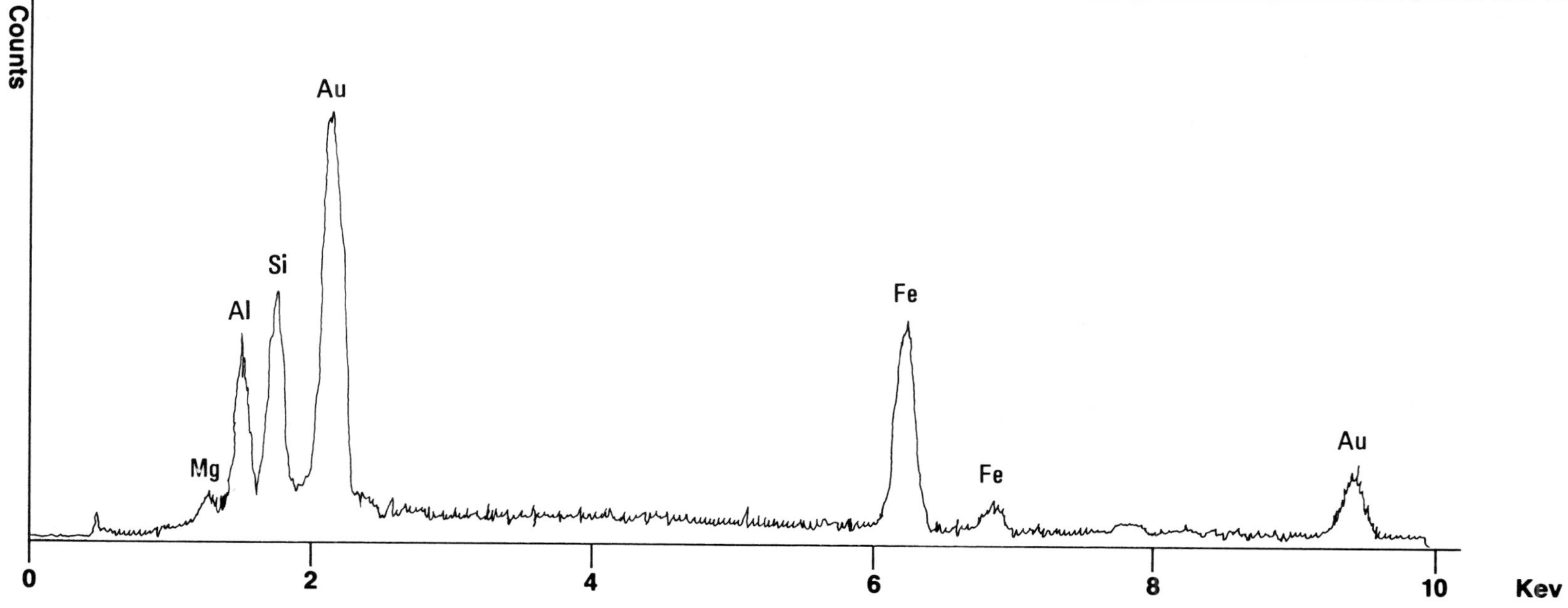

Silicates—Clay (Chlorite)

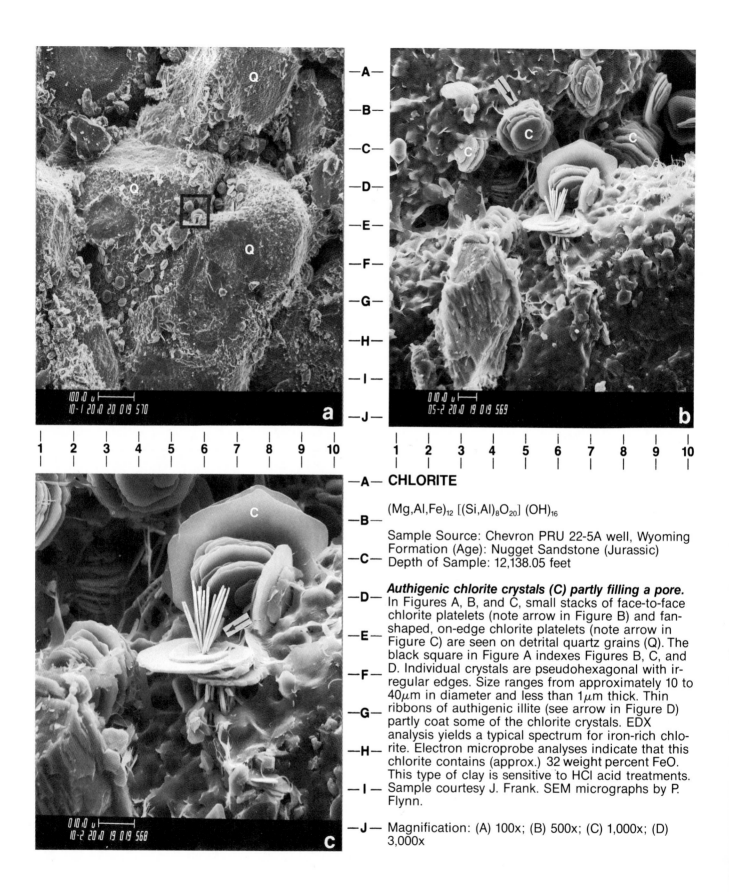

—A— CHLORITE

$(Mg,Al,Fe)_{12} [(Si,Al)_8 O_{20}] (OH)_{16}$

Sample Source: Chevron PRU 22-5A well, Wyoming
Formation (Age): Nugget Sandstone (Jurassic)
Depth of Sample: 12,138.05 feet

Authigenic chlorite crystals (C) partly filling a pore.
In Figures A, B, and C, small stacks of face-to-face
chlorite platelets (note arrow in Figure B) and fan-
shaped, on-edge chlorite platelets (note arrow in
Figure C) are seen on detrital quartz grains (Q). The
black square in Figure A indexes Figures B, C, and
D. Individual crystals are pseudohexagonal with irregular edges. Size ranges from approximately 10 to
$40\mu m$ in diameter and less than $1\mu m$ thick. Thin
ribbons of authigenic illite (see arrow in Figure D)
partly coat some of the chlorite crystals. EDX
analysis yields a typical spectrum for iron-rich chlorite. Electron microprobe analyses indicate that this
chlorite contains (approx.) 32 weight percent FeO.
This type of clay is sensitive to HCl acid treatments.
Sample courtesy J. Frank. SEM micrographs by P.
Flynn.

Magnification: (A) 100x; (B) 500x; (C) 1,000x; (D)
3,000x

Silicates—Clay (Chlorite)

Energy Dispersive X—Ray Spectrum (EDX)

Chlorite $(Mg, Al, Fe)_{12} [(Si, Al)_8 O_{20}] (OH)_{16}$

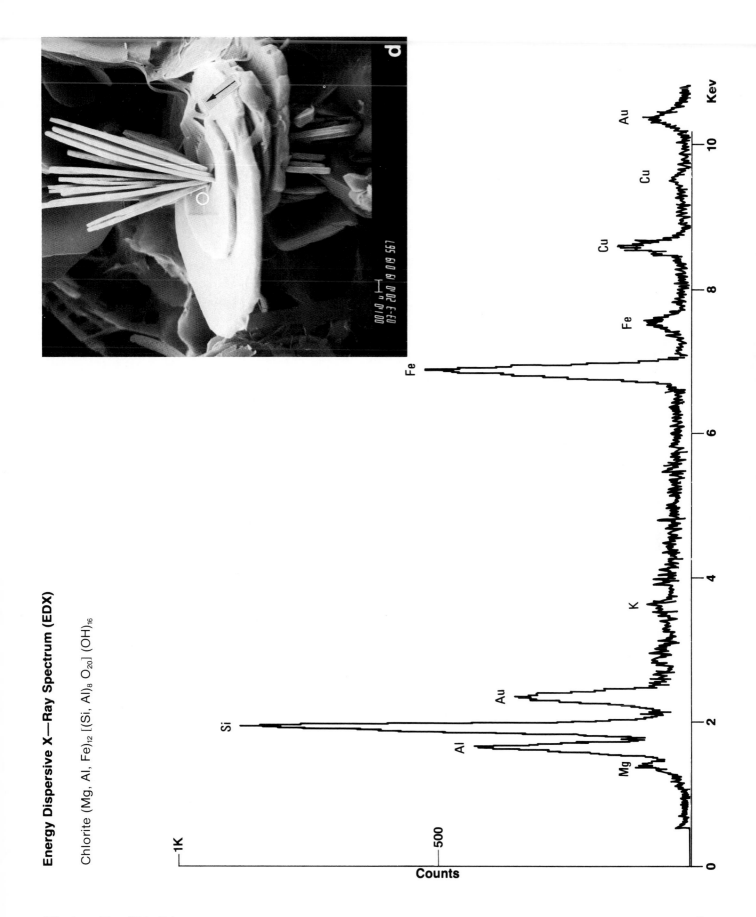

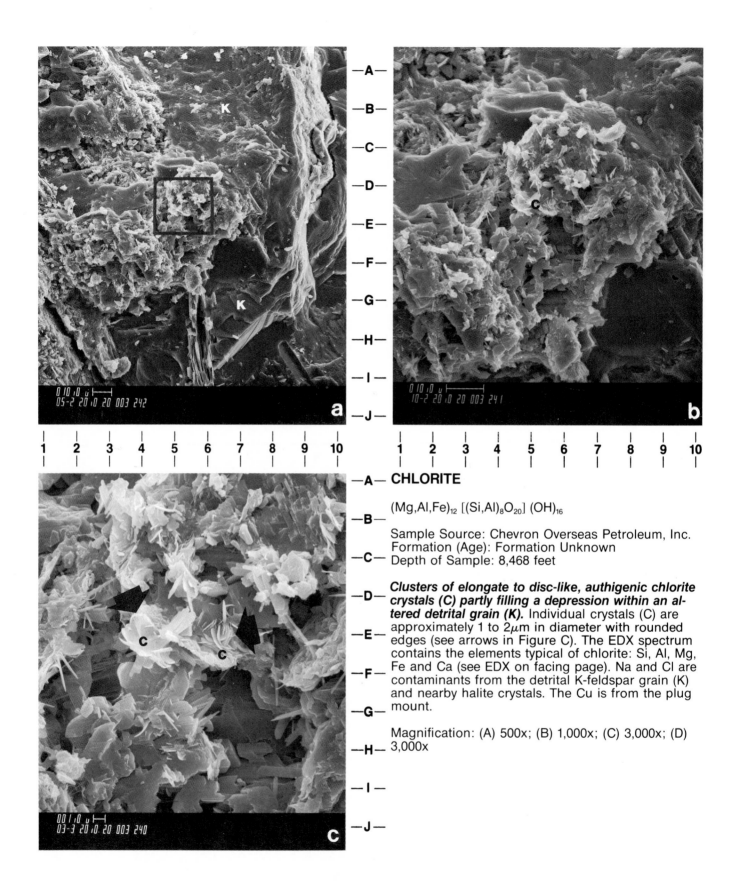

CHLORITE

$(Mg,Al,Fe)_{12} [(Si,Al)_8 O_{20}] (OH)_{16}$

Sample Source: Chevron Overseas Petroleum, Inc.
Formation (Age): Formation Unknown
Depth of Sample: 8,468 feet

Clusters of elongate to disc-like, authigenic chlorite crystals (C) partly filling a depression within an altered detrital grain (K). Individual crystals (C) are approximately 1 to 2μm in diameter with rounded edges (see arrows in Figure C). The EDX spectrum contains the elements typical of chlorite: Si, Al, Mg, Fe and Ca (see EDX on facing page). Na and Cl are contaminants from the detrital K-feldspar grain (K) and nearby halite crystals. The Cu is from the plug mount.

Magnification: (A) 500x; (B) 1,000x; (C) 3,000x; (D) 3,000x

Energy Dispersive X—Ray Spectrum (EDX)

Chlorite (Mg, Al, Fe)$_{12}$ [(Si, Al)$_8$ O$_{20}$] (OH)$_{16}$

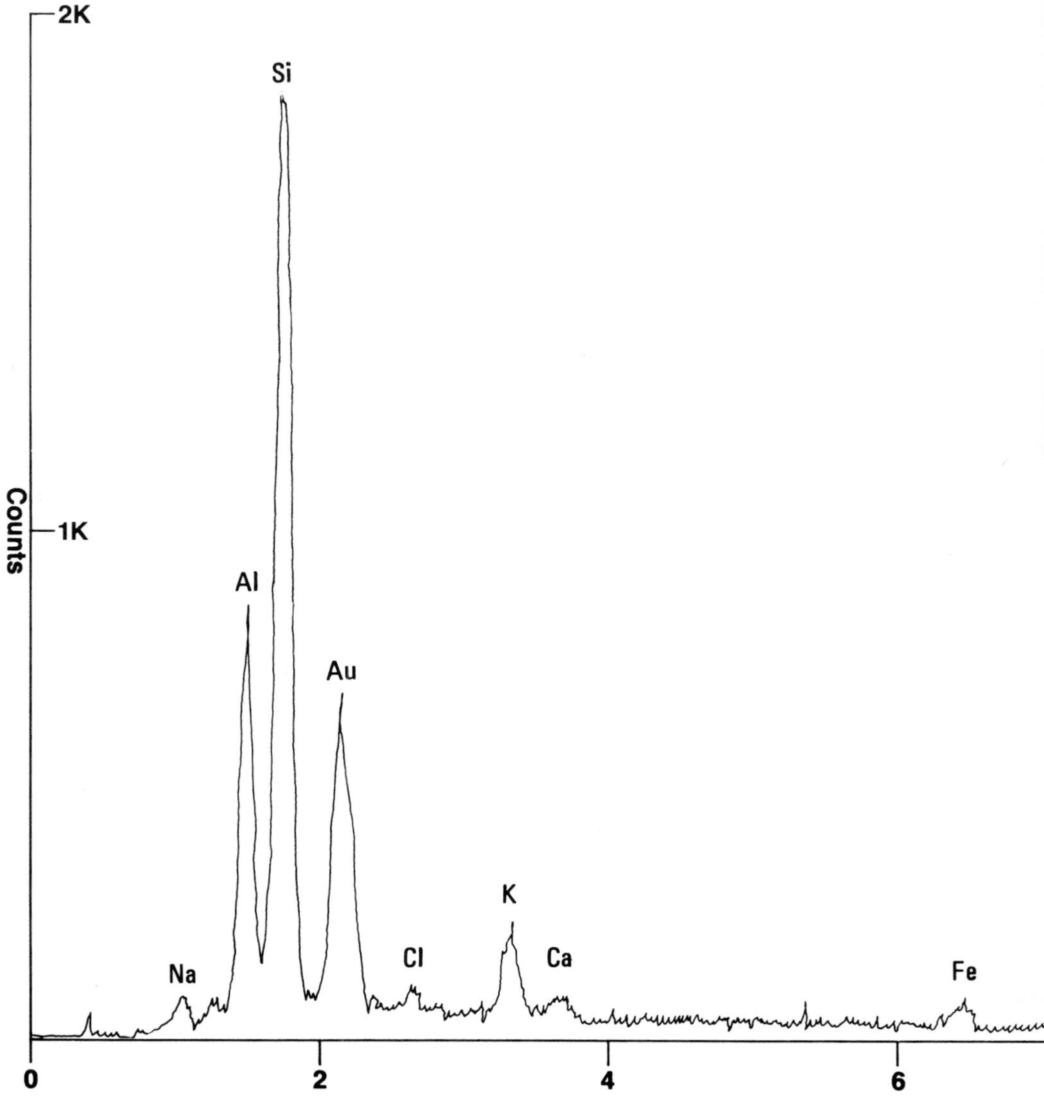

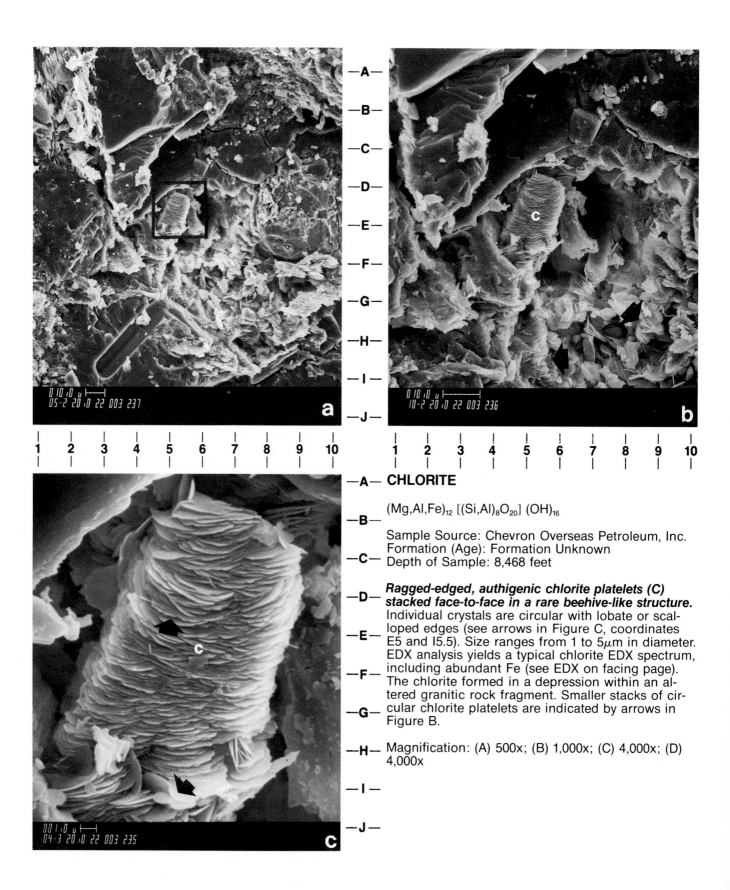

CHLORITE

$(Mg,Al,Fe)_{12} [(Si,Al)_8O_{20}] (OH)_{16}$

Sample Source: Chevron Overseas Petroleum, Inc.
Formation (Age): Formation Unknown
Depth of Sample: 8,468 feet

Ragged-edged, authigenic chlorite platelets (C) stacked face-to-face in a rare beehive-like structure. Individual crystals are circular with lobate or scalloped edges (see arrows in Figure C, coordinates E5 and I5.5). Size ranges from 1 to $5\mu m$ in diameter. EDX analysis yields a typical chlorite EDX spectrum, including abundant Fe (see EDX on facing page). The chlorite formed in a depression within an altered granitic rock fragment. Smaller stacks of circular chlorite platelets are indicated by arrows in Figure B.

Magnification: (A) 500x; (B) 1,000x; (C) 4,000x; (D) 4,000x

Silicates—Clay (Chlorite)

Energy Dispersive X—Ray Spectrum (EDX)

Chlorite $(Mg, Al, Fe)_{12} [(Si, Al)_8 O_{20}] (OH)_{16}$

Silicates—Clay (Chlorite)

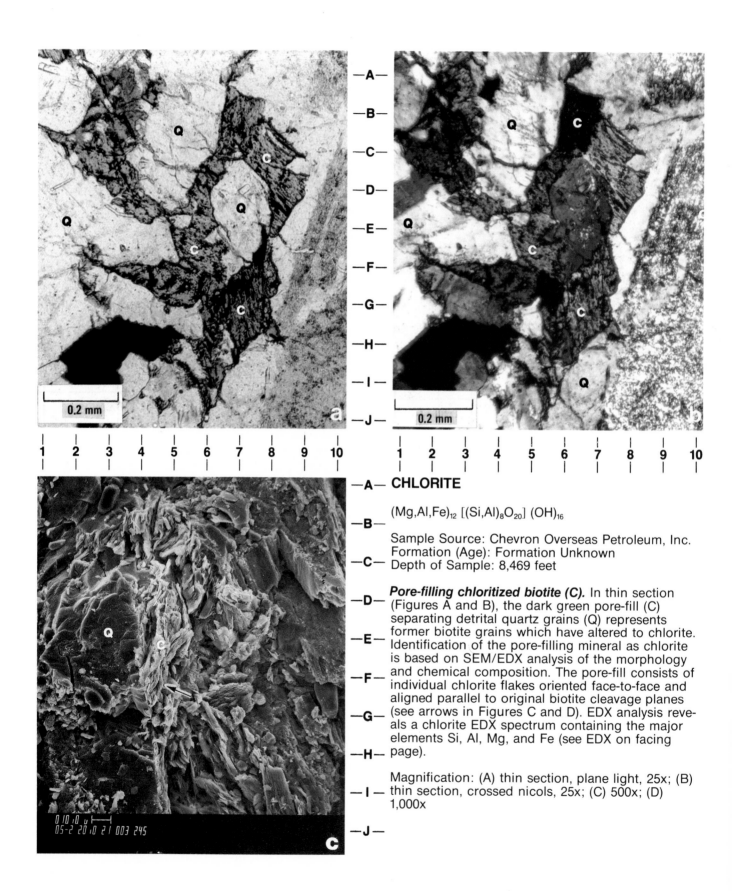

—A— CHLORITE

$(Mg,Al,Fe)_{12} [(Si,Al)_8 O_{20}] (OH)_{16}$

Sample Source: Chevron Overseas Petroleum, Inc.
Formation (Age): Formation Unknown
Depth of Sample: 8,469 feet

Pore-filling chloritized biotite (C). In thin section (Figures A and B), the dark green pore-fill (C) separating detrital quartz grains (Q) represents former biotite grains which have altered to chlorite. Identification of the pore-filling mineral as chlorite is based on SEM/EDX analysis of the morphology and chemical composition. The pore-fill consists of individual chlorite flakes oriented face-to-face and aligned parallel to original biotite cleavage planes (see arrows in Figures C and D). EDX analysis reveals a chlorite EDX spectrum containing the major elements Si, Al, Mg, and Fe (see EDX on facing page).

Magnification: (A) thin section, plane light, 25x; (B) thin section, crossed nicols, 25x; (C) 500x; (D) 1,000x

Silicates—Clay (Chlorite)

Energy Dispersive X—Ray Spectrum (EDX)

Chlorite $(Mg, Al, Fe)_{12}$ $[(Si, Al)_8 O_{20}]$ $(OH)_{16}$

Silicates—Clay (Chlorite)

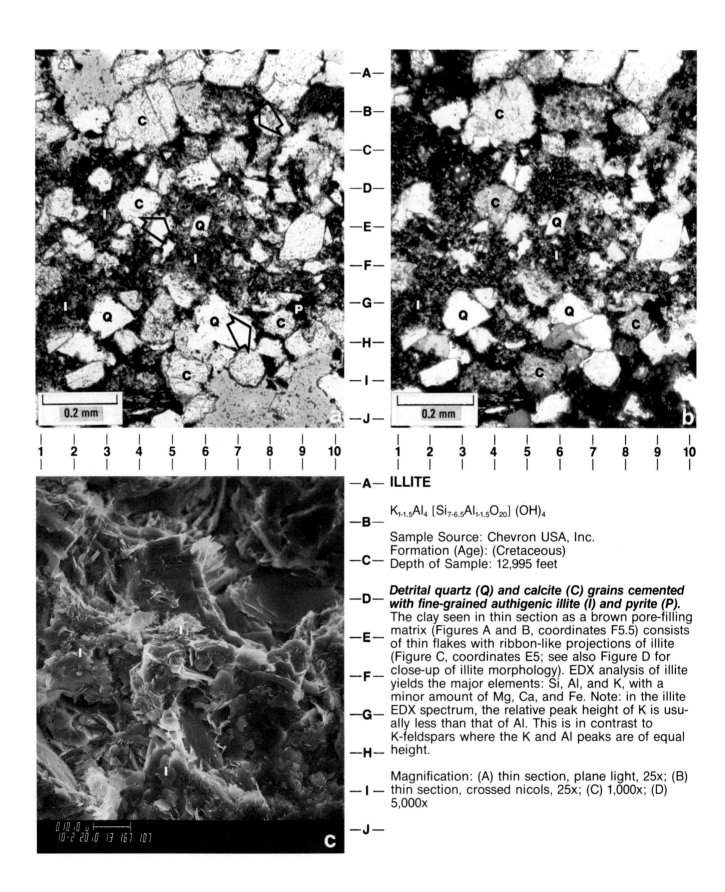

ILLITE

$K_{1-1.5}Al_4 [Si_{7-6.5}Al_{1-1.5}O_{20}] (OH)_4$

Sample Source: Chevron USA, Inc.
Formation (Age): (Cretaceous)
Depth of Sample: 12,995 feet

Detrital quartz (Q) and calcite (C) grains cemented with fine-grained authigenic illite (I) and pyrite (P). The clay seen in thin section as a brown pore-filling matrix (Figures A and B, coordinates F5.5) consists of thin flakes with ribbon-like projections of illite (Figure C, coordinates E5; see also Figure D for close-up of illite morphology). EDX analysis of illite yields the major elements: Si, Al, and K, with a minor amount of Mg, Ca, and Fe. Note: in the illite EDX spectrum, the relative peak height of K is usually less than that of Al. This is in contrast to K-feldspars where the K and Al peaks are of equal height.

Magnification: (A) thin section, plane light, 25x; (B) thin section, crossed nicols, 25x; (C) 1,000x; (D) 5,000x

Energy Dispersive X—Ray Spectrum (EDX)

Illite $K_{1-1.5}$ Al_4 $[Si_{7-6.5}$ $Al_{1-1.5}$ $O_{20}]$ $(OH)_4$

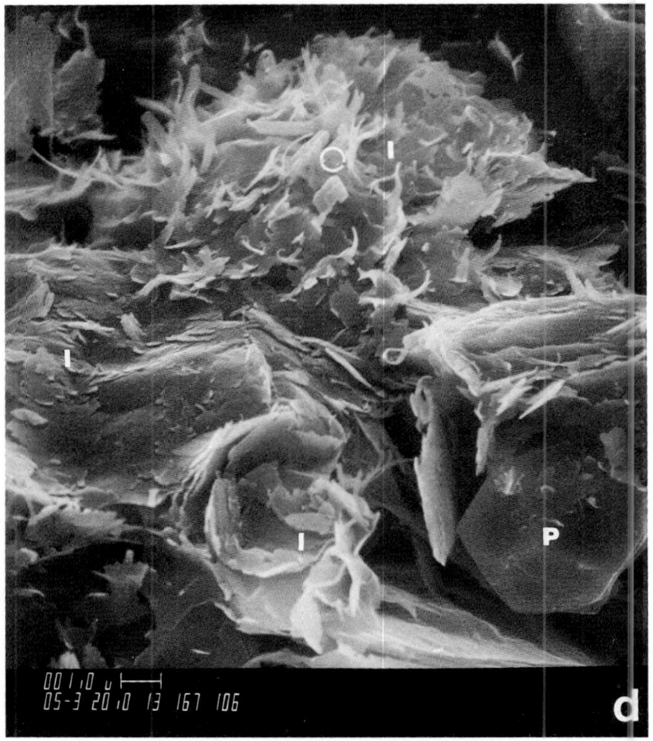

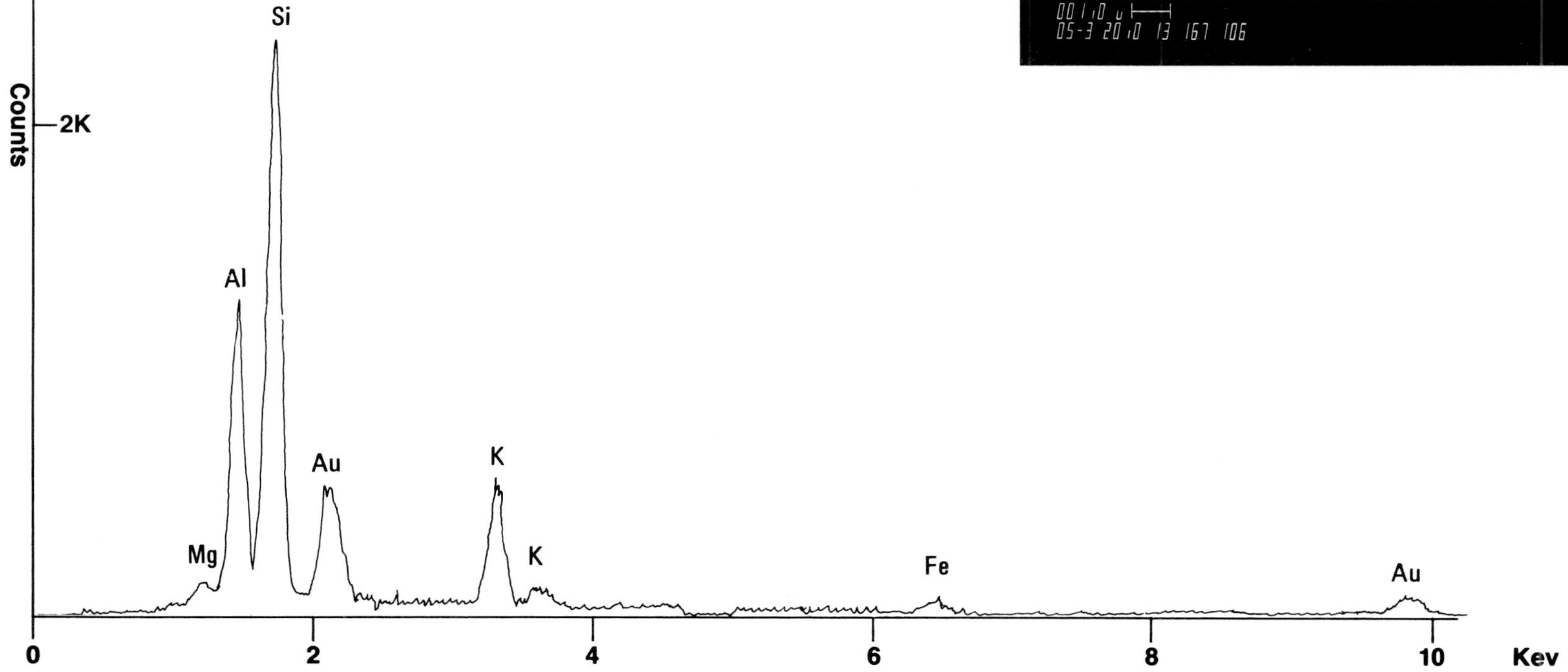

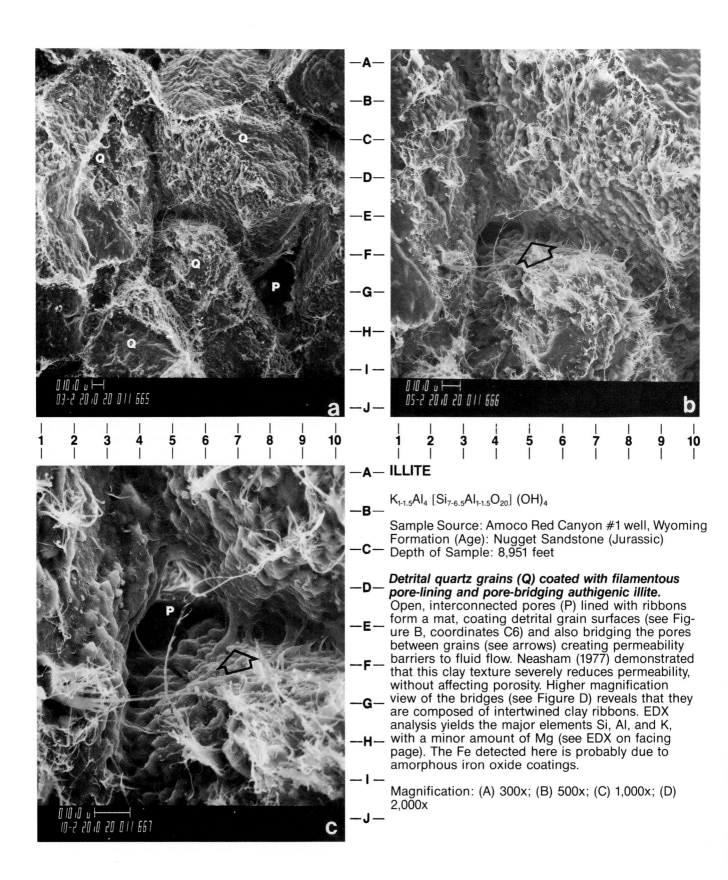

ILLITE

$K_{1-1.5}Al_4 [Si_{7-6.5}Al_{1-1.5}O_{20}] (OH)_4$

Sample Source: Amoco Red Canyon #1 well, Wyoming
Formation (Age): Nugget Sandstone (Jurassic)
Depth of Sample: 8,951 feet

Detrital quartz grains (Q) coated with filamentous pore-lining and pore-bridging authigenic illite.
Open, interconnected pores (P) lined with ribbons form a mat, coating detrital grain surfaces (see Figure B, coordinates C6) and also bridging the pores between grains (see arrows) creating permeability barriers to fluid flow. Neasham (1977) demonstrated that this clay texture severely reduces permeability, without affecting porosity. Higher magnification view of the bridges (see Figure D) reveals that they are composed of intertwined clay ribbons. EDX analysis yields the major elements Si, Al, and K, with a minor amount of Mg (see EDX on facing page). The Fe detected here is probably due to amorphous iron oxide coatings.

Magnification: (A) 300x; (B) 500x; (C) 1,000x; (D) 2,000x

Energy Dispersive X—Ray Spectrum (EDX)

Illite $K_{1-1.5}$ Al_4 $[Si_{7-6.5}$ $Al_{1-1.5}$ $O_{20}]$ $(OH)_4$

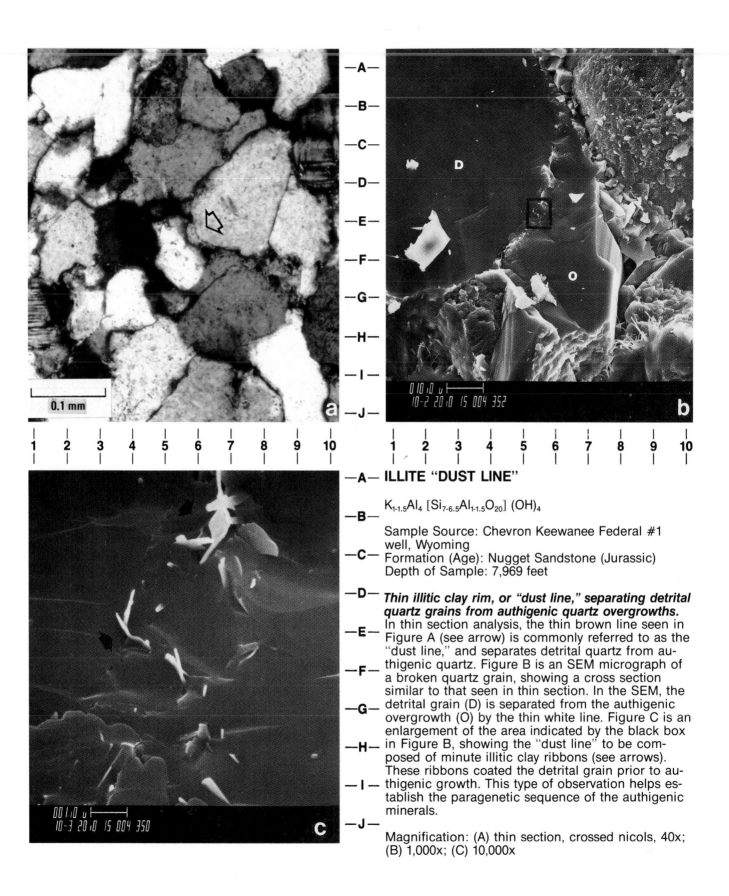

ILLITE "DUST LINE"

$K_{1-1.5}Al_4 [Si_{7-6.5}Al_{1-1.5}O_{20}] (OH)_4$

Sample Source: Chevron Keewanee Federal #1 well, Wyoming
Formation (Age): Nugget Sandstone (Jurassic)
Depth of Sample: 7,969 feet

Thin illitic clay rim, or "dust line," separating detrital quartz grains from authigenic quartz overgrowths.
In thin section analysis, the thin brown line seen in Figure A (see arrow) is commonly referred to as the "dust line," and separates detrital quartz from authigenic quartz. Figure B is an SEM micrograph of a broken quartz grain, showing a cross section similar to that seen in thin section. In the SEM, the detrital grain (D) is separated from the authigenic overgrowth (O) by the thin white line. Figure C is an enlargement of the area indicated by the black box in Figure B, showing the "dust line" to be composed of minute illitic clay ribbons (see arrows). These ribbons coated the detrital grain prior to authigenic growth. This type of observation helps establish the paragenetic sequence of the authigenic minerals.

Magnification: (A) thin section, crossed nicols, 40x; (B) 1,000x; (C) 10,000x

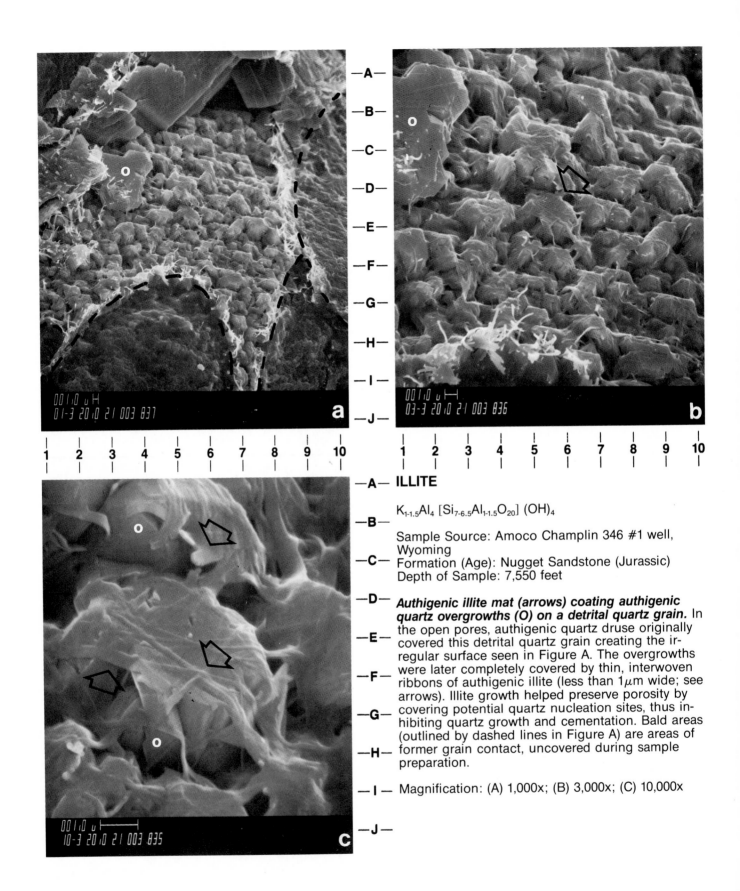

ILLITE

$K_{1-1.5}Al_4 [Si_{7-6.5}Al_{1-1.5}O_{20}] (OH)_4$

Sample Source: Amoco Champlin 346 #1 well, Wyoming
Formation (Age): Nugget Sandstone (Jurassic)
Depth of Sample: 7,550 feet

Authigenic illite mat (arrows) coating authigenic quartz overgrowths (O) on a detrital quartz grain. In the open pores, authigenic quartz druse originally covered this detrital quartz grain creating the irregular surface seen in Figure A. The overgrowths were later completely covered by thin, interwoven ribbons of authigenic illite (less than $1\mu m$ wide; see arrows). Illite growth helped preserve porosity by covering potential quartz nucleation sites, thus inhibiting quartz growth and cementation. Bald areas (outlined by dashed lines in Figure A) are areas of former grain contact, uncovered during sample preparation.

Magnification: (A) 1,000x; (B) 3,000x; (C) 10,000x

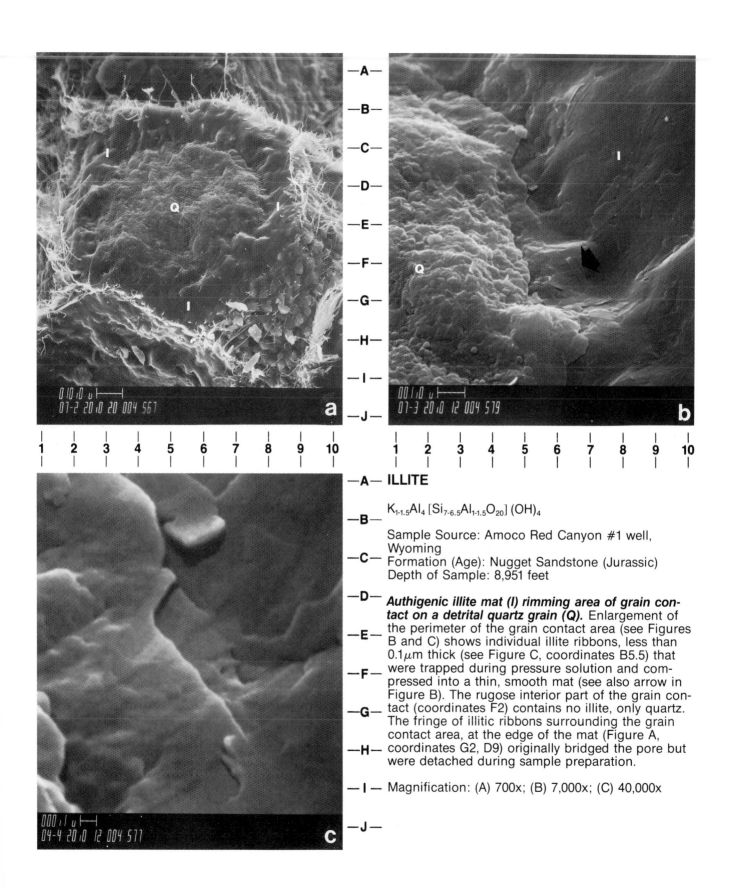

ILLITE

$K_{1-1.5}Al_4 [Si_{7-6.5}Al_{1-1.5}O_{20}] (OH)_4$

Sample Source: Amoco Red Canyon #1 well, Wyoming
Formation (Age): Nugget Sandstone (Jurassic)
Depth of Sample: 8,951 feet

__Authigenic illite mat (I) rimming area of grain contact on a detrital quartz grain (Q).__ Enlargement of the perimeter of the grain contact area (see Figures B and C) shows individual illite ribbons, less than 0.1µm thick (see Figure C, coordinates B5.5) that were trapped during pressure solution and compressed into a thin, smooth mat (see also arrow in Figure B). The rugose interior part of the grain contact (coordinates F2) contains no illite, only quartz. The fringe of illitic ribbons surrounding the grain contact area, at the edge of the mat (Figure A, coordinates G2, D9) originally bridged the pore but were detached during sample preparation.

Magnification: (A) 700x; (B) 7,000x; (C) 40,000x

ILLITE

$K_{1-1.5}Al_4 [Si_{7-6.5}Al_{1-1.5}O_{20}] (OH)_4$

Sample Source: API Reference Clay #36, Morris, Illinois
Formation (Age): (Pennsylvanian)
Depth of Sample: Open Pit Mine

Massive detrital illite composed of irregular, flake-like clay platelets oriented parallel to each other.
The flaky morphology of this detrital illite is not unique to illite, thus is of no help in identifying this clay. Precise identification is based on X-ray diffraction (XRD) analysis and supported by EDX analysis (see EDX on facing page). The EDX spectrum is similar to those figured for the other authigenic illites in this atlas (except for the presence of Ti). Note: although this morphology is not diagnostic of illites, it is a common morphology of detrital clays and can be helpful in separating detrital and authigenic clays.

Magnification: (A) 500x; (B) 5,000x; (C) 10,000x; (D) 20,000x

Energy Dispersive X—Ray Spectrum (EDX)

Illite $K_{1-1.5} Al_4 [Si_{7-6.5} Al_{1-1.5} O_{20}] (OH)_4$

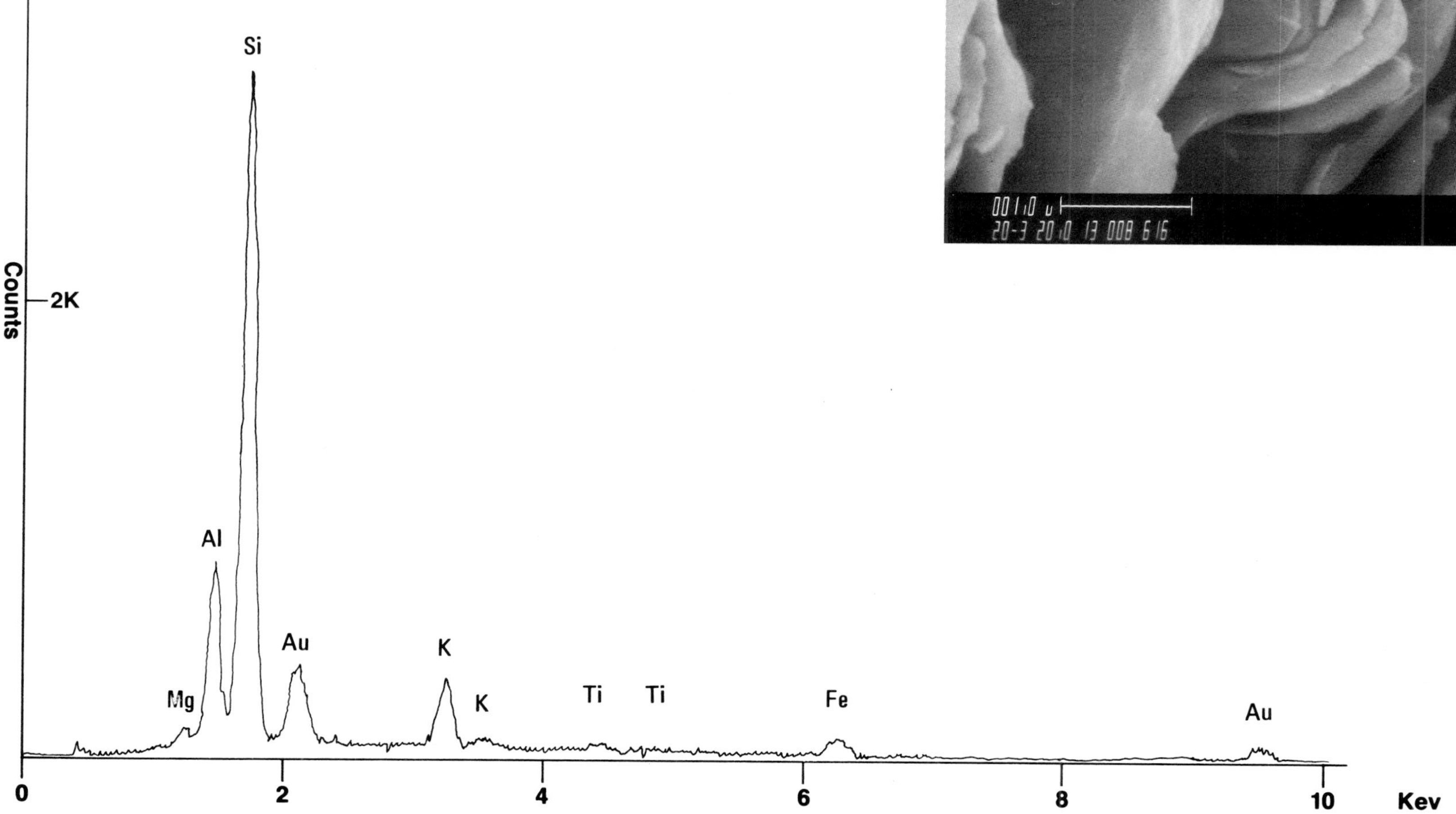

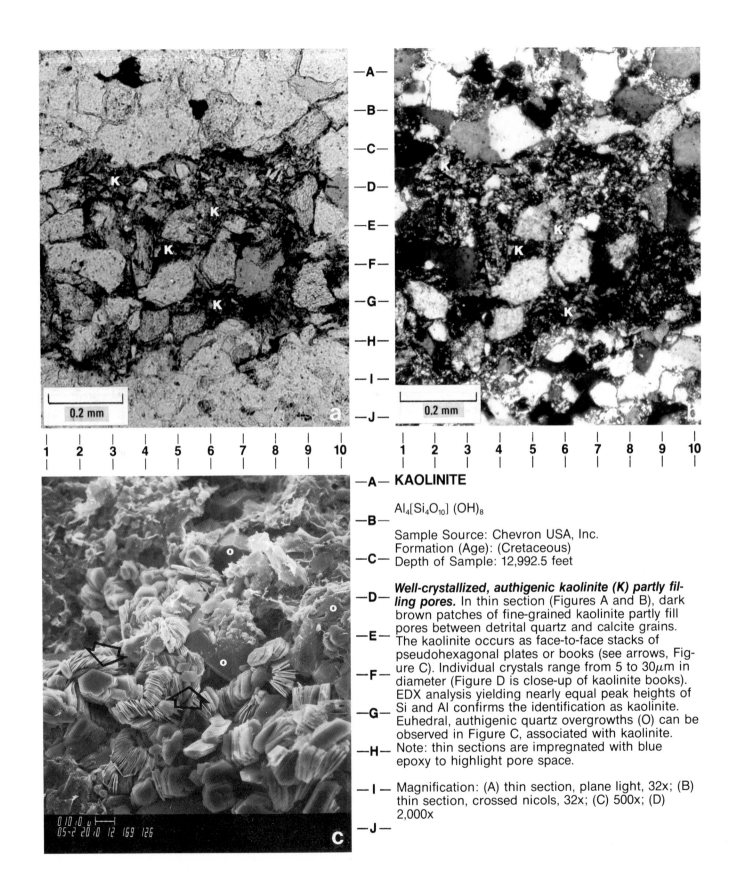

KAOLINITE

$Al_4[Si_4O_{10}](OH)_8$

Sample Source: Chevron USA, Inc.
Formation (Age): (Cretaceous)
Depth of Sample: 12,992.5 feet

Well-crystallized, authigenic kaolinite (K) partly filling pores. In thin section (Figures A and B), dark brown patches of fine-grained kaolinite partly fill pores between detrital quartz and calcite grains. The kaolinite occurs as face-to-face stacks of pseudohexagonal plates or books (see arrows, Figure C). Individual crystals range from 5 to 30μm in diameter (Figure D is close-up of kaolinite books). EDX analysis yielding nearly equal peak heights of Si and Al confirms the identification as kaolinite. Euhedral, authigenic quartz overgrowths (O) can be observed in Figure C, associated with kaolinite. Note: thin sections are impregnated with blue epoxy to highlight pore space.

Magnification: (A) thin section, plane light, 32x; (B) thin section, crossed nicols, 32x; (C) 500x; (D) 2,000x

Silicates—Clay (Kaolinite)

Energy Dispersive X—Ray Spectrum (EDX)

Kaolinite $Al_4 [Si_4 O_{10}] (OH)_8$

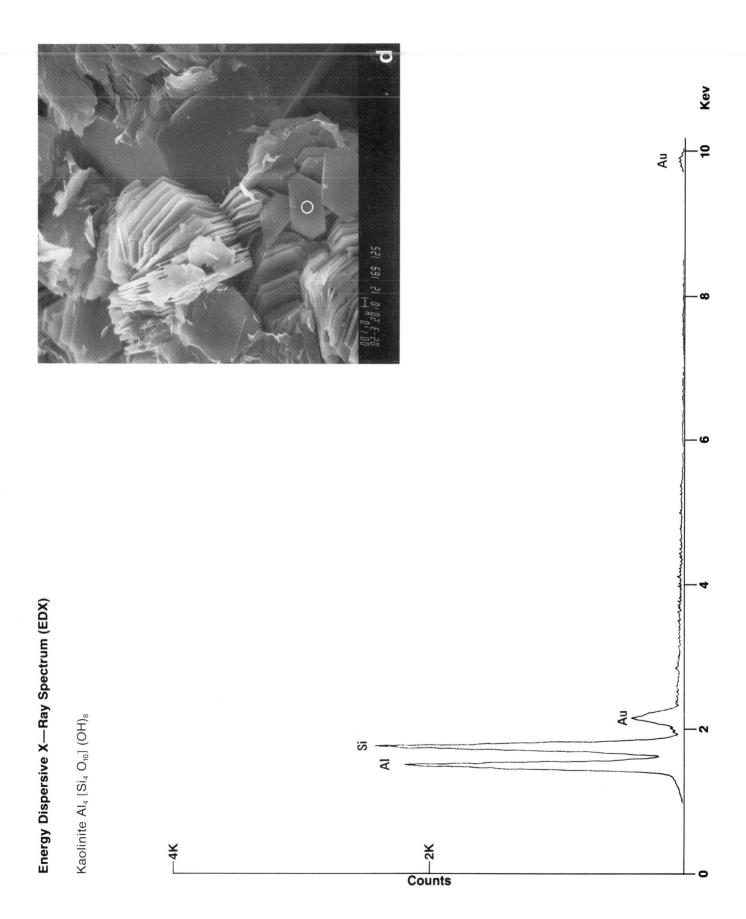

KAOLINITE

$Al_4[Si_4O_{10}] (OH)_8$

Sample Source: Chevron Overseas Petroleum, Inc.
Formation (Age): (Jurassic)
Depth of Sample: 10,456.5 feet

Blocky, pore-filling kaolinite (K). This example
shows books of kaolinite, approximately 1 μm thick
(see Figures B and C, coordinates E5), separated
and linked by ribbons or sheets of filamentous
illite-smectite (see arrows, Figures C and D) partly filling a pore. Figure D is a close-up view of the individual books. The blocky morphology and typical
kaolinite EDX spectrum are diagnostic (see EDX on
facing page). Micropores are visible within the pore
filling (see Figure B, coordinates C6, G9; Figure C,
coordinates E7).

Magnification: (A) 500x; (B) 1,000x; (C) 2,000x; (D)
10,000x

Silicates—Clay (Kaolinite)

Energy Dispersive X—Ray Spectrum (EDX)

Kaolinite $Al_4 [Si_4 O_{10}] (OH)_8$

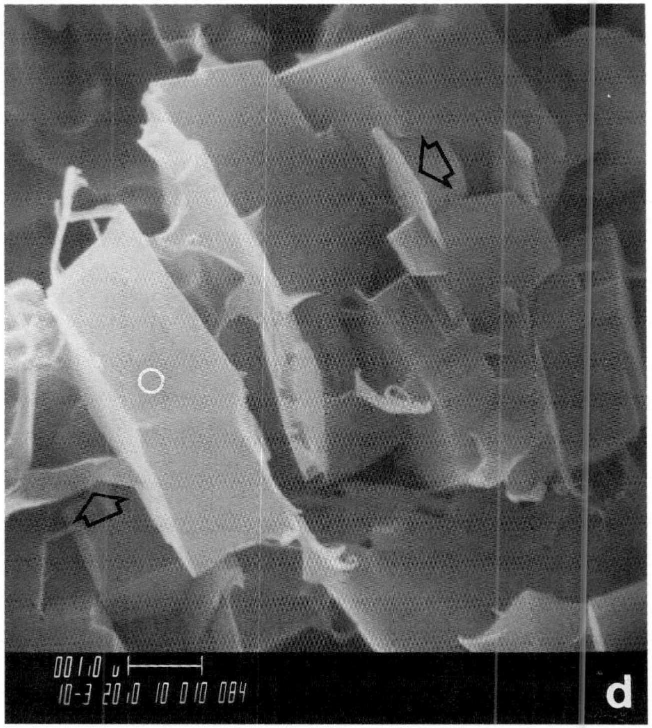

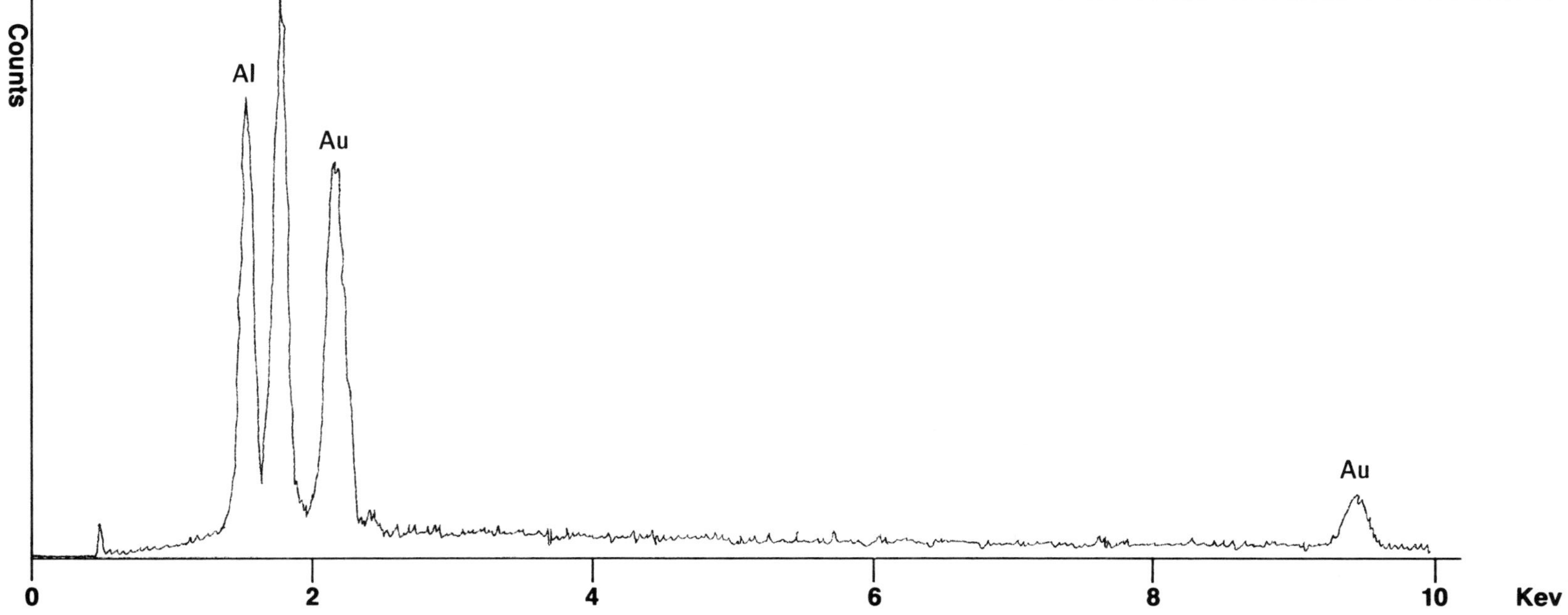

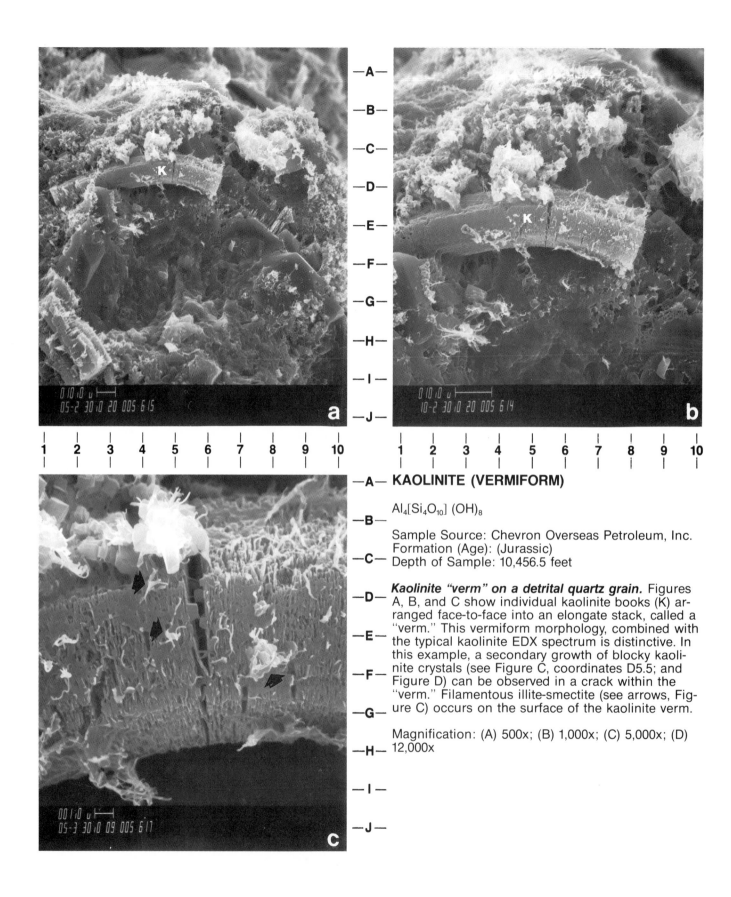

KAOLINITE (VERMIFORM)

$Al_4[Si_4O_{10}] (OH)_8$

Sample Source: Chevron Overseas Petroleum, Inc.
Formation (Age): (Jurassic)
Depth of Sample: 10,456.5 feet

Kaolinite "verm" on a detrital quartz grain. Figures A, B, and C show individual kaolinite books (K) arranged face-to-face into an elongate stack, called a "verm." This vermiform morphology, combined with the typical kaolinite EDX spectrum is distinctive. In this example, a secondary growth of blocky kaolinite crystals (see Figure C, coordinates D5.5; and Figure D) can be observed in a crack within the "verm." Filamentous illite-smectite (see arrows, Figure C) occurs on the surface of the kaolinite verm.

Magnification: (A) 500x; (B) 1,000x; (C) 5,000x; (D) 12,000x

Energy Dispersive X—Ray Spectrum (EDX)

Kaolinite $Al_4 [Si_4 O_{10}] (OH)_8$

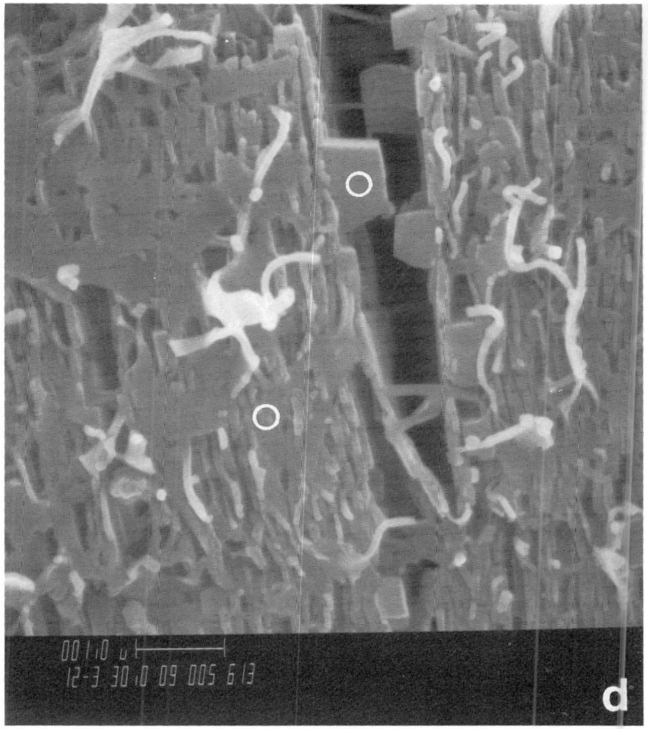

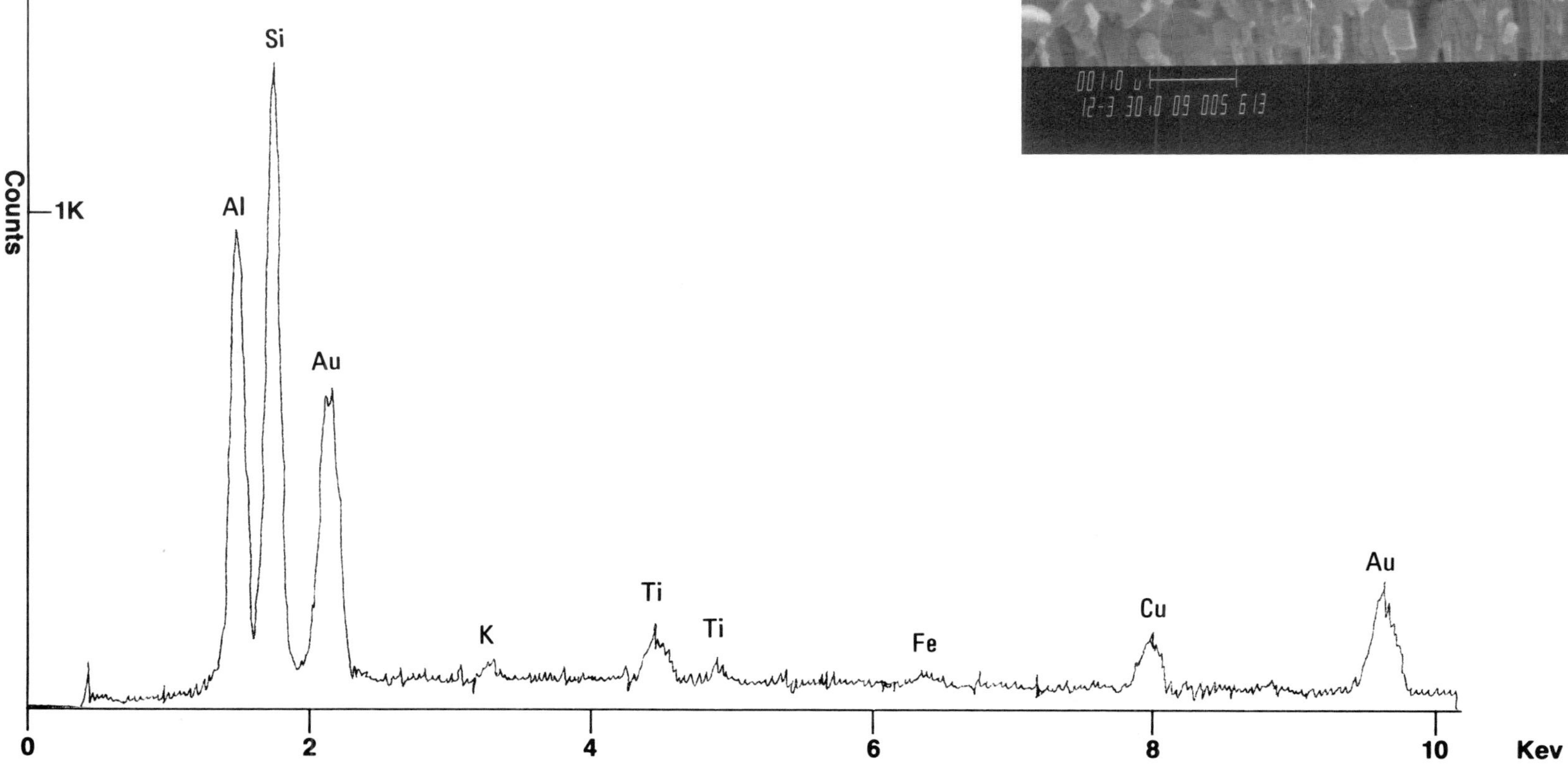

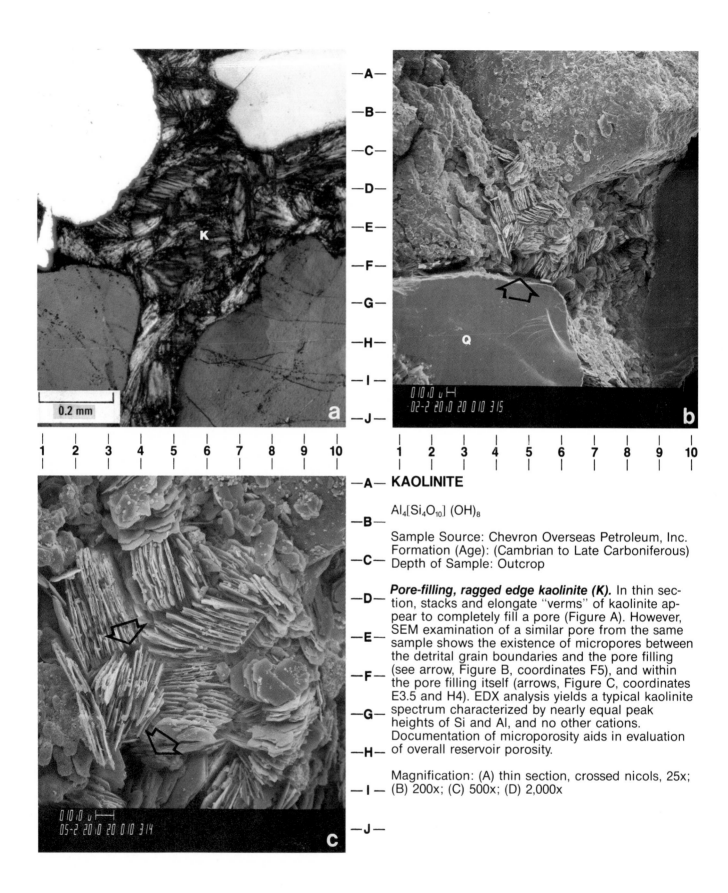

KAOLINITE

$Al_4[Si_4O_{10}](OH)_8$

Sample Source: Chevron Overseas Petroleum, Inc.
Formation (Age): (Cambrian to Late Carboniferous)
Depth of Sample: Outcrop

Pore-filling, ragged edge kaolinite (K). In thin section, stacks and elongate "verms" of kaolinite appear to completely fill a pore (Figure A). However, SEM examination of a similar pore from the same sample shows the existence of micropores between the detrital grain boundaries and the pore filling (see arrow, Figure B, coordinates F5), and within the pore filling itself (arrows, Figure C, coordinates E3.5 and H4). EDX analysis yields a typical kaolinite spectrum characterized by nearly equal peak heights of Si and Al, and no other cations. Documentation of microporosity aids in evaluation of overall reservoir porosity.

Magnification: (A) thin section, crossed nicols, 25x; (B) 200x; (C) 500x; (D) 2,000x

Silicates—Clay (Kaolinite)

Energy Dispersive X—Ray Spectrum (EDX)

Kaolinite $Al_4 [Si_4 O_{10}] (OH)_8$

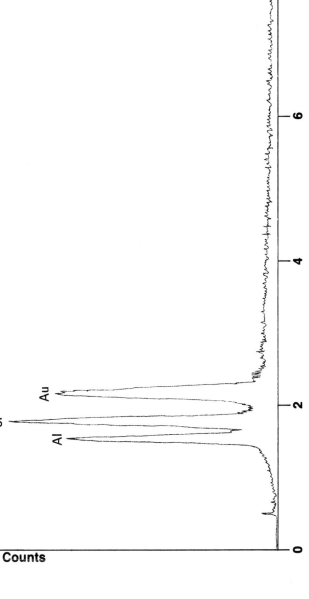

2K

Counts

Al
Si
Au

Au

Kev

0
2
4
6
8
10

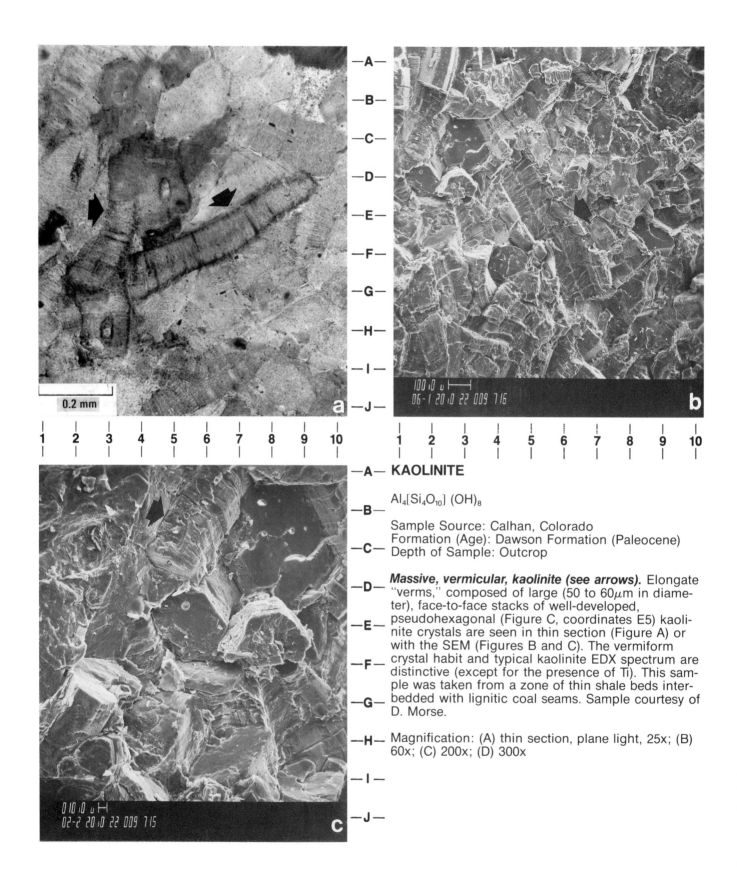

KAOLINITE

$Al_4[Si_4O_{10}](OH)_8$

Sample Source: Calhan, Colorado
Formation (Age): Dawson Formation (Paleocene)
Depth of Sample: Outcrop

Massive, vermicular, kaolinite (see arrows). Elongate "verms," composed of large (50 to 60μm in diameter), face-to-face stacks of well-developed, pseudohexagonal (Figure C, coordinates E5) kaolinite crystals are seen in thin section (Figure A) or with the SEM (Figures B and C). The vermiform crystal habit and typical kaolinite EDX spectrum are distinctive (except for the presence of Ti). This sample was taken from a zone of thin shale beds interbedded with lignitic coal seams. Sample courtesy of D. Morse.

Magnification: (A) thin section, plane light, 25x; (B) 60x; (C) 200x; (D) 300x

Energy Dispersive X—Ray Spectrum (EDX)

Kaolinite $Al_4 [Si_4 O_{10}] (OH)_8$

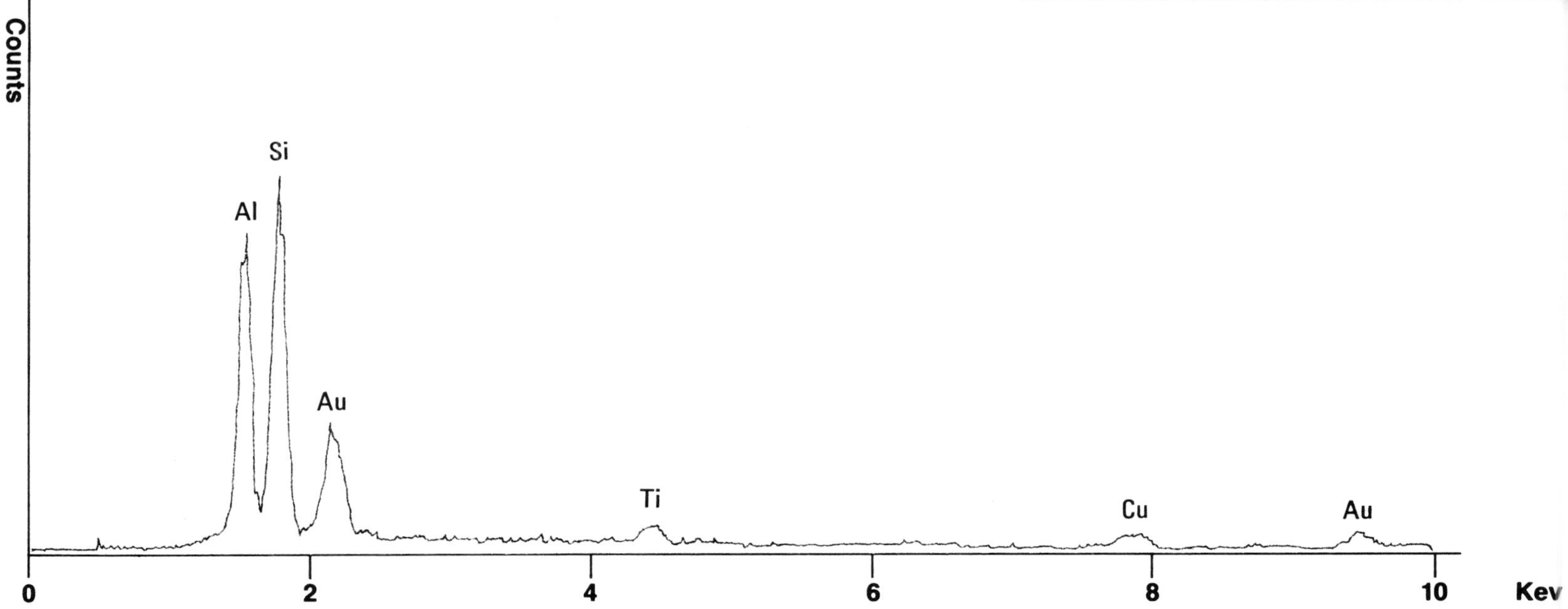

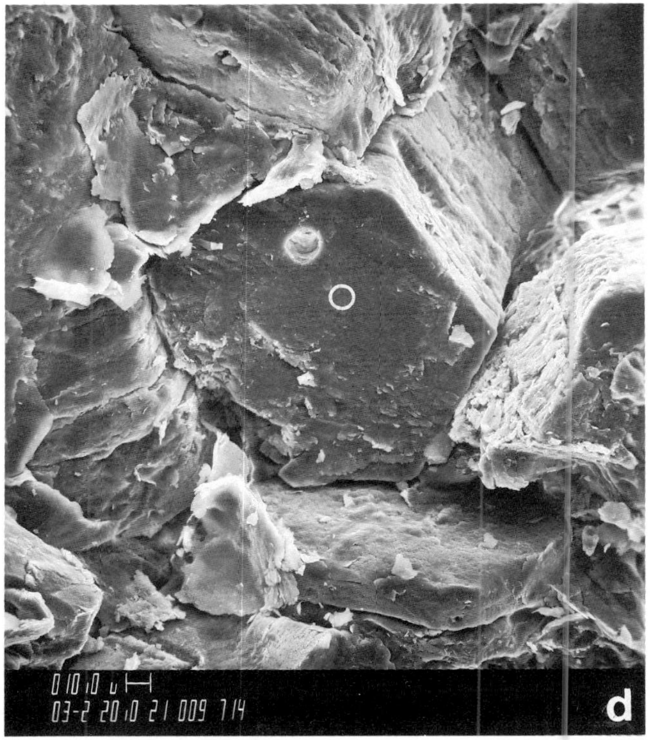

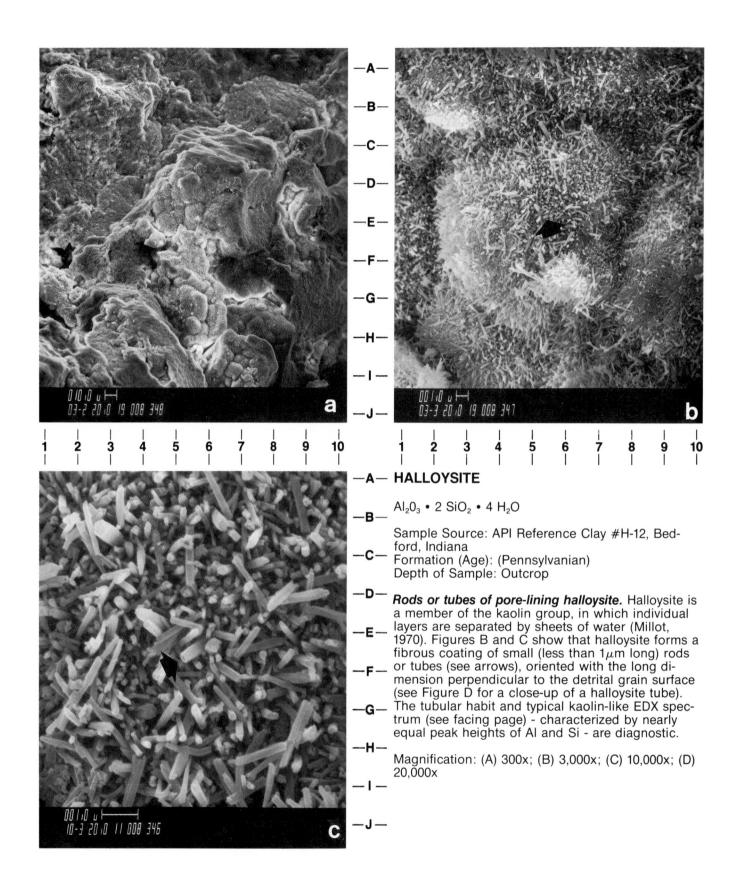

—A— **HALLOYSITE**

—B— $Al_2O_3 \cdot 2\ SiO_2 \cdot 4\ H_2O$

—C— Sample Source: API Reference Clay #H-12, Bedford, Indiana
Formation (Age): (Pennsylvanian)
Depth of Sample: Outcrop

—D— ***Rods or tubes of pore-lining halloysite.*** Halloysite is a member of the kaolin group, in which individual layers are separated by sheets of water (Millot, 1970). Figures B and C show that halloysite forms a fibrous coating of small (less than $1\mu m$ long) rods or tubes (see arrows), oriented with the long dimension perpendicular to the detrital grain surface (see Figure D for a close-up of a halloysite tube). The tubular habit and typical kaolin-like EDX spectrum (see facing page) - characterized by nearly equal peak heights of Al and Si - are diagnostic.

Magnification: (A) 300x; (B) 3,000x; (C) 10,000x; (D) 20,000x

Energy Dispersive X—Ray Spectrum (EDX)

Halloysite $Al_2O_3 \cdot 2\ SiO_2 \cdot 4\ H_2O$

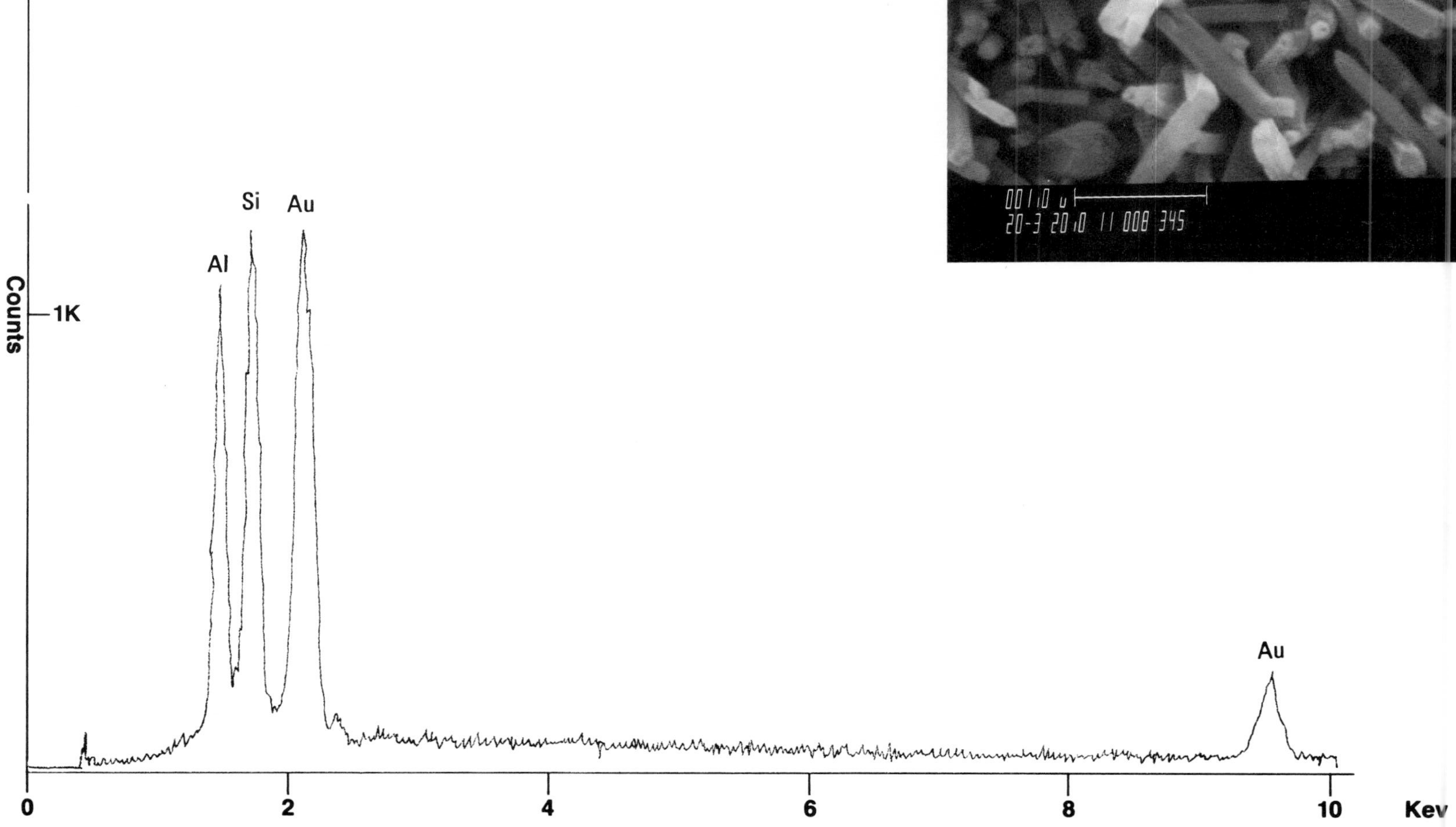

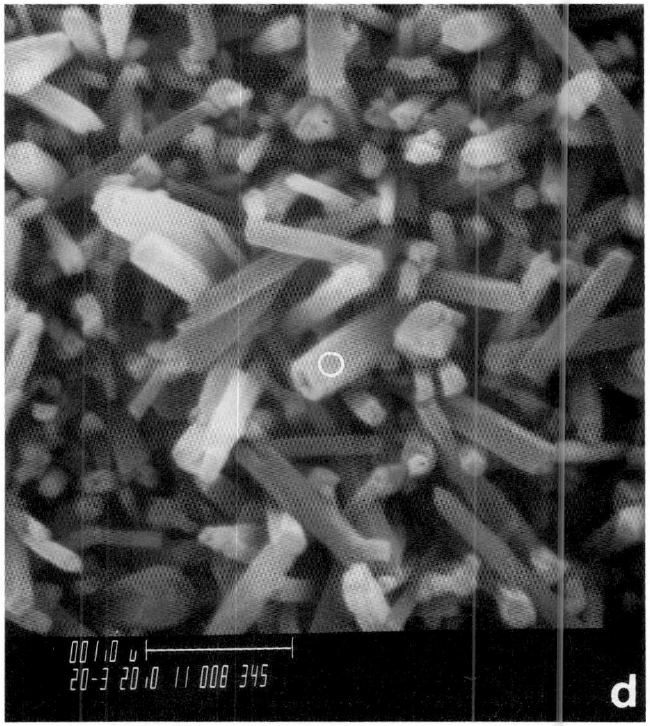

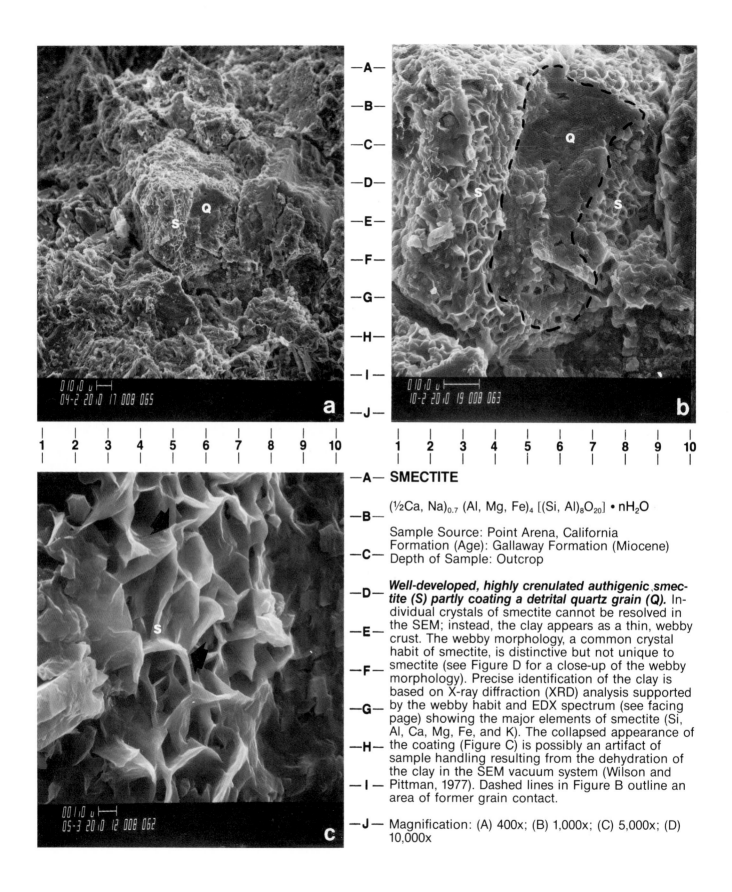

A—

B— **SMECTITE**

$(\frac{1}{2}Ca, Na)_{0.7} (Al, Mg, Fe)_4 [(Si, Al)_8O_{20}] \cdot nH_2O$

Sample Source: Point Arena, California
Formation (Age): Gallaway Formation (Miocene)
Depth of Sample: Outcrop

Well-developed, highly crenulated authigenic smectite (S) partly coating a detrital quartz grain (Q). Individual crystals of smectite cannot be resolved in the SEM; instead, the clay appears as a thin, webby crust. The webby morphology, a common crystal habit of smectite, is distinctive but not unique to smectite (see Figure D for a close-up of the webby morphology). Precise identification of the clay is based on X-ray diffraction (XRD) analysis supported by the webby habit and EDX spectrum (see facing page) showing the major elements of smectite (Si, Al, Ca, Mg, Fe, and K). The collapsed appearance of the coating (Figure C) is possibly an artifact of sample handling resulting from the dehydration of the clay in the SEM vacuum system (Wilson and Pittman, 1977). Dashed lines in Figure B outline an area of former grain contact.

Magnification: (A) 400x; (B) 1,000x; (C) 5,000x; (D) 10,000x

Silicates—Clay (Smectite)

Energy Dispersive X—Ray Spectrum (EDX)

Smectite (½Ca, Na)$_{0.7}$ (Al, Mg, Fe)$_4$ [(Si, Al)$_8$O$_{20}$] • nH$_2$O

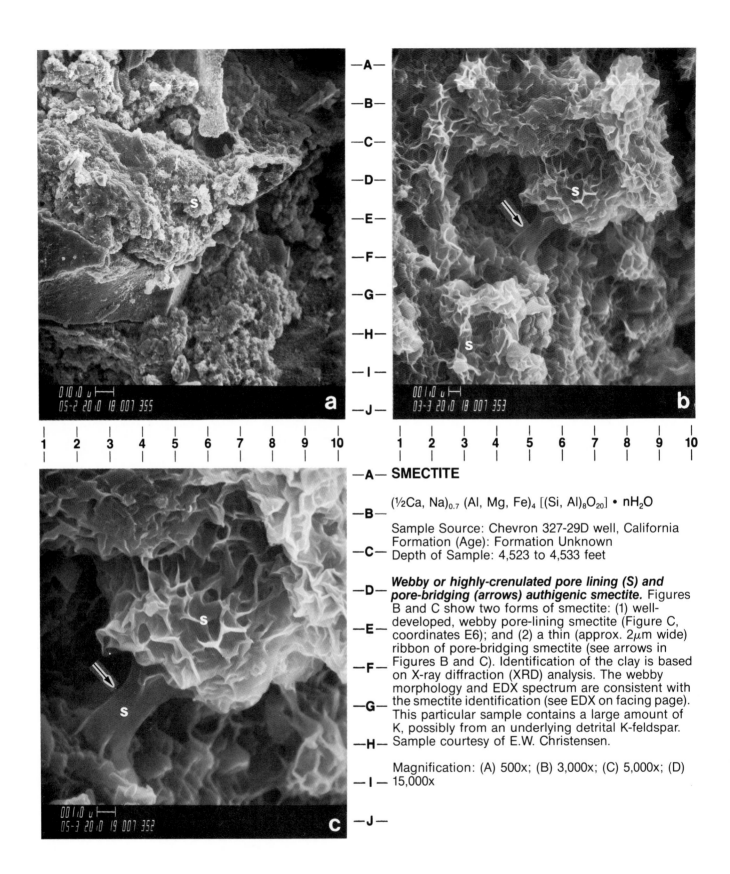

1 2 3 4 5 6 7 8 9 10

1 2 3 4 5 6 7 8 9 10

—A— **SMECTITE**

—B— $(\frac{1}{2}Ca, Na)_{0.7} (Al, Mg, Fe)_4 [(Si, Al)_8O_{20}] \cdot nH_2O$

—C—
Sample Source: Chevron 327-29D well, California
Formation (Age): Formation Unknown
Depth of Sample: 4,523 to 4,533 feet

—D—
***Webby or highly-crenulated pore lining (S) and
pore-bridging (arrows) authigenic smectite.*** Figures
B and C show two forms of smectite: (1) well-
developed, webby pore-lining smectite (Figure C,
—E— coordinates E6); and (2) a thin (approx. 2μm wide)
ribbon of pore-bridging smectite (see arrows in
Figures B and C). Identification of the clay is based
—F— on X-ray diffraction (XRD) analysis. The webby
morphology and EDX spectrum are consistent with
the smectite identification (see EDX on facing page).
—G— This particular sample contains a large amount of
K, possibly from an underlying detrital K-feldspar.
—H— Sample courtesy of E.W. Christensen.

Magnification: (A) 500x; (B) 3,000x; (C) 5,000x; (D)
—I— 15,000x

—J—

Energy Dispersive X—Ray Spectrum (EDX)

Smectite (½Ca, Na)$_{0.7}$ (Al, Mg, Fe)$_4$ [(Si, Al)$_8$O$_{20}$] • nH$_2$O

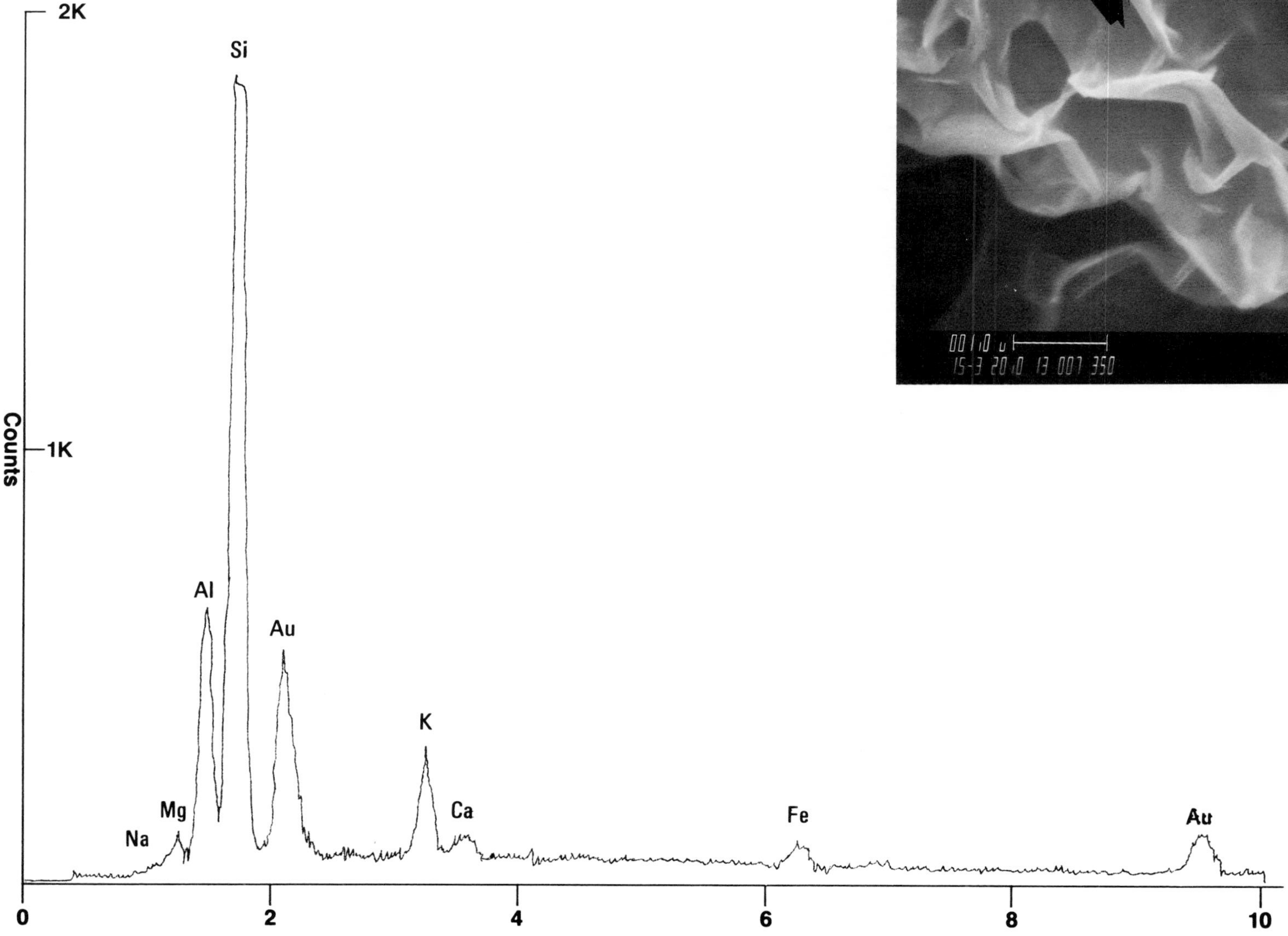

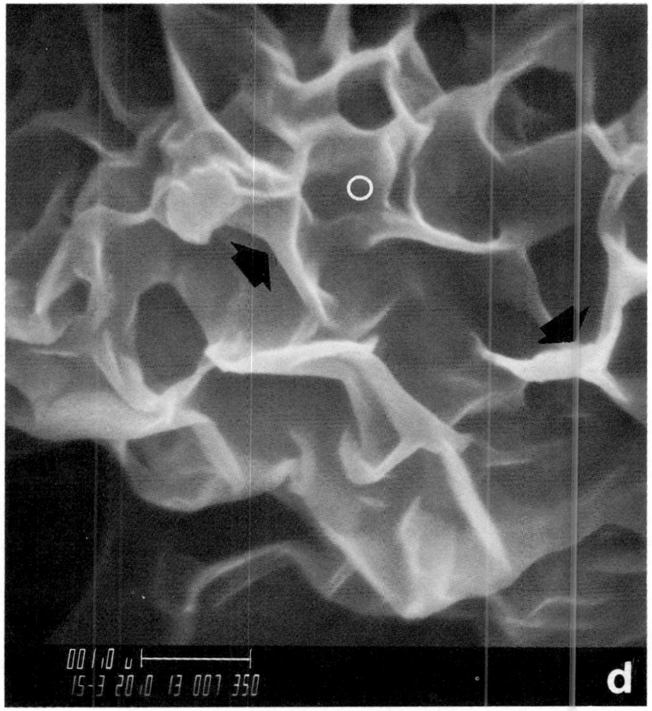

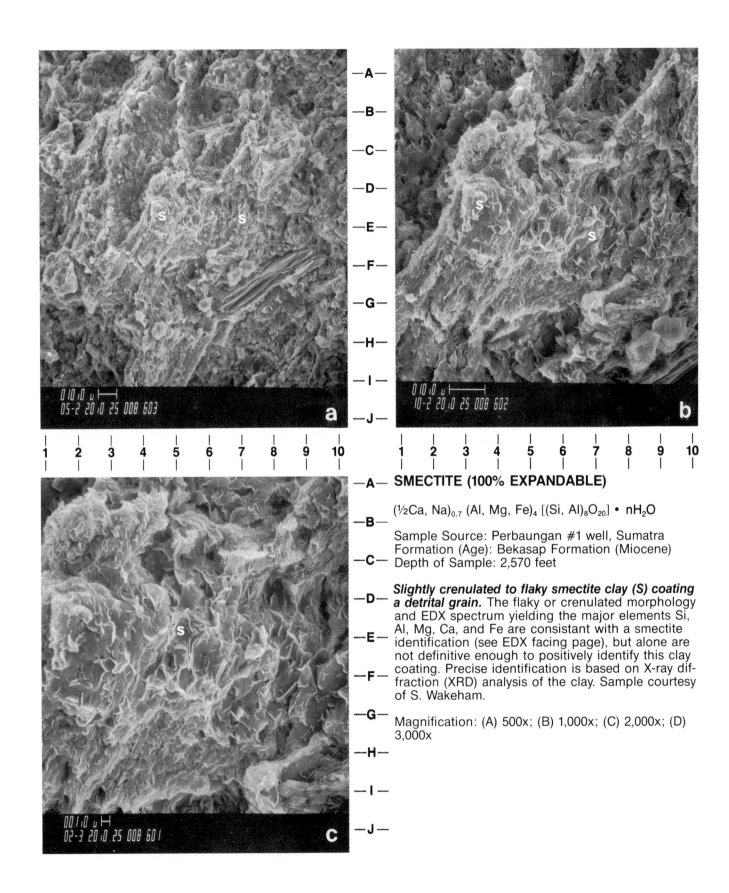

1 2 3 4 5 6 7 8 9 10

SMECTITE (100% EXPANDABLE)

$(\frac{1}{2}Ca, Na)_{0.7} (Al, Mg, Fe)_4 [(Si, Al)_8O_{20}] \cdot nH_2O$

Sample Source: Perbaungan #1 well, Sumatra
Formation (Age): Bekasap Formation (Miocene)
Depth of Sample: 2,570 feet

***Slightly crenulated to flaky smectite clay (S) coating
a detrital grain.*** The flaky or crenulated morphology
and EDX spectrum yielding the major elements Si,
Al, Mg, Ca, and Fe are consistant with a smectite
identification (see EDX facing page), but alone are
not definitive enough to positively identify this clay
coating. Precise identification is based on X-ray diffraction (XRD) analysis of the clay. Sample courtesy
of S. Wakeham.

Magnification: (A) 500x; (B) 1,000x; (C) 2,000x; (D)
3,000x

Energy Dispersive X—Ray Spectrum (EDX)

Smectite (½Ca, Na)$_{0.7}$ (Al, Mg, Fe)$_4$ [(Si, Al)$_8$O$_{20}$] • nH$_2$O

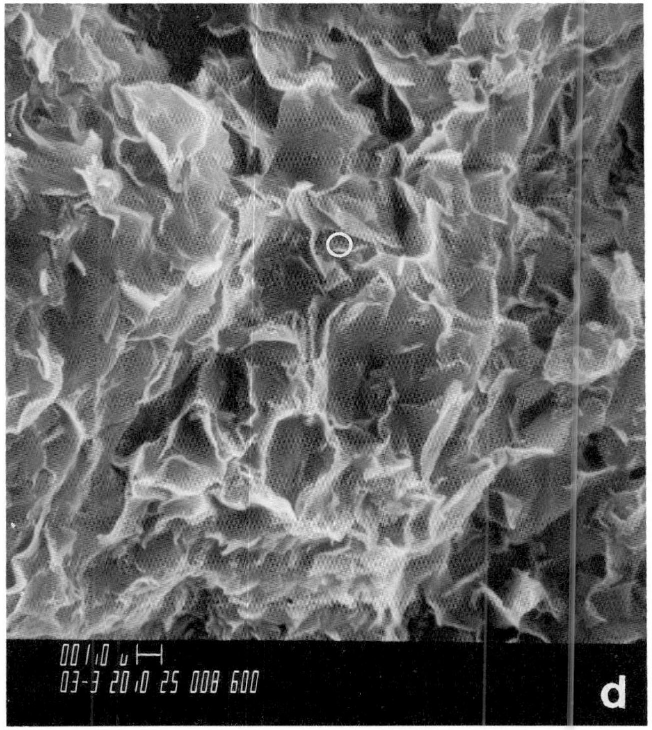

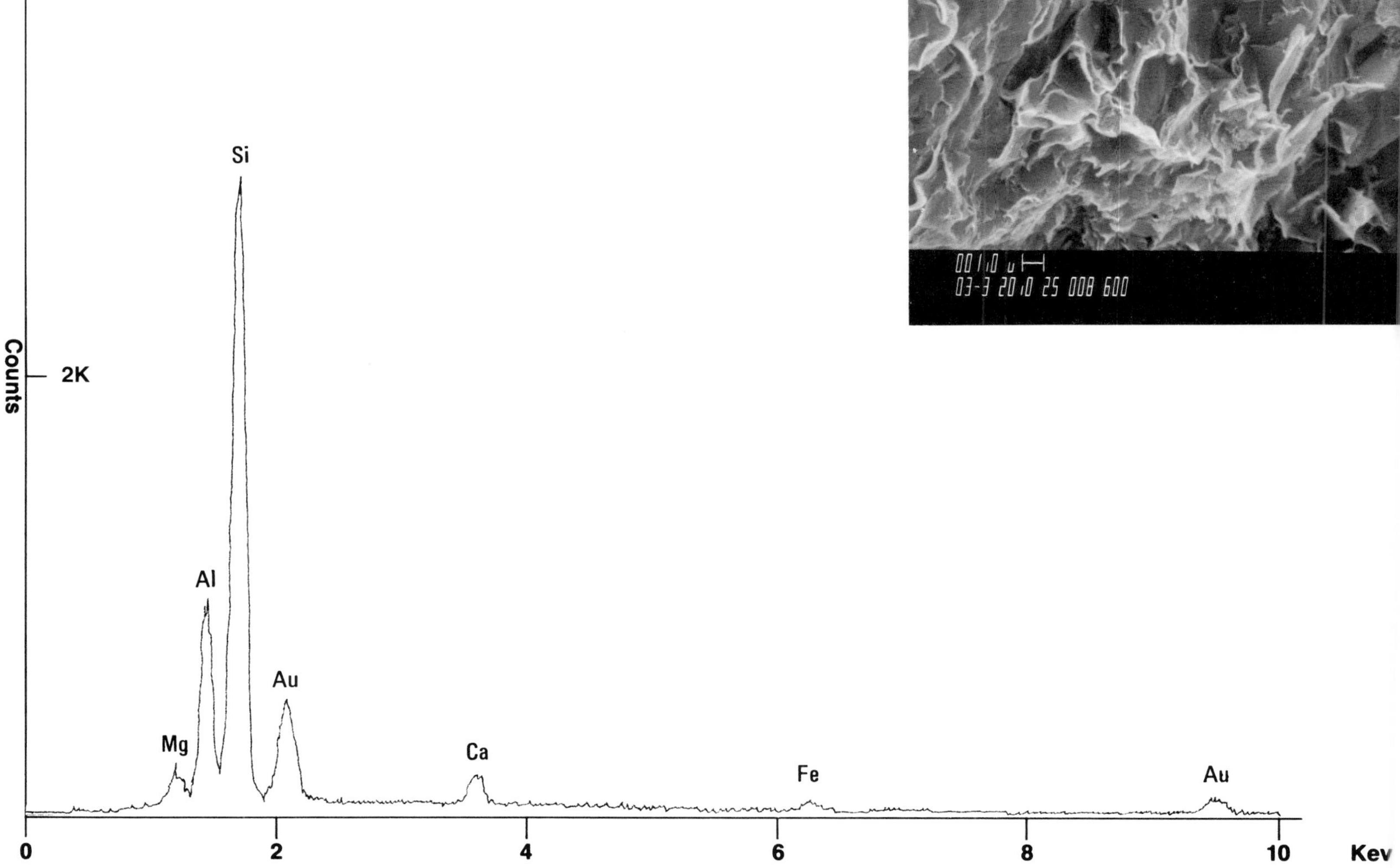

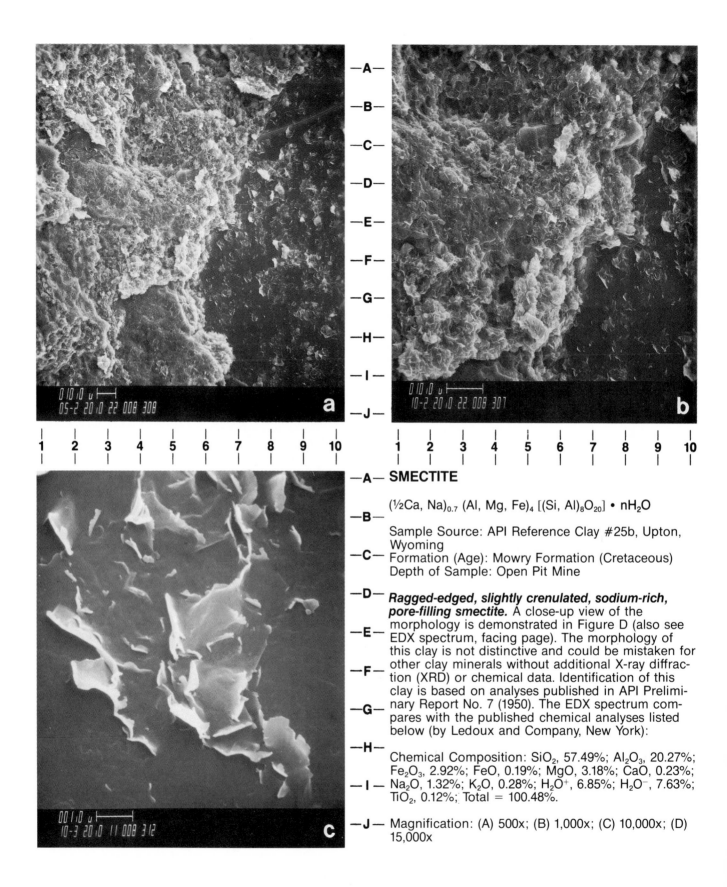

SMECTITE

$(\frac{1}{2}Ca, Na)_{0.7} (Al, Mg, Fe)_4 [(Si, Al)_8O_{20}] \cdot nH_2O$

Sample Source: API Reference Clay #25b, Upton, Wyoming
Formation (Age): Mowry Formation (Cretaceous)
Depth of Sample: Open Pit Mine

Ragged-edged, slightly crenulated, sodium-rich, pore-filling smectite. A close-up view of the morphology is demonstrated in Figure D (also see EDX spectrum, facing page). The morphology of this clay is not distinctive and could be mistaken for other clay minerals without additional X-ray diffraction (XRD) or chemical data. Identification of this clay is based on analyses published in API Preliminary Report No. 7 (1950). The EDX spectrum compares with the published chemical analyses listed below (by Ledoux and Company, New York):

Chemical Composition: SiO_2, 57.49%; Al_2O_3, 20.27%; Fe_2O_3, 2.92%; FeO, 0.19%; MgO, 3.18%; CaO, 0.23%; Na_2O, 1.32%; K_2O, 0.28%; H_2O^+, 6.85%; H_2O^-, 7.63%; TiO_2, 0.12%; Total = 100.48%.

Magnification: (A) 500x; (B) 1,000x; (C) 10,000x; (D) 15,000x

Energy Dispersive X—Ray Spectrum (EDX)

Smectite ($\frac{1}{2}$Ca, Na)$_{0.7}$ (Al, Mg, Fe)$_4$ [(Si, Al)$_8$O$_{20}$] • nH$_2$O

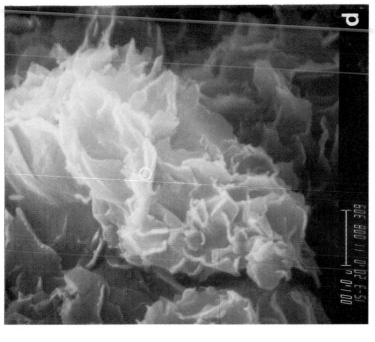

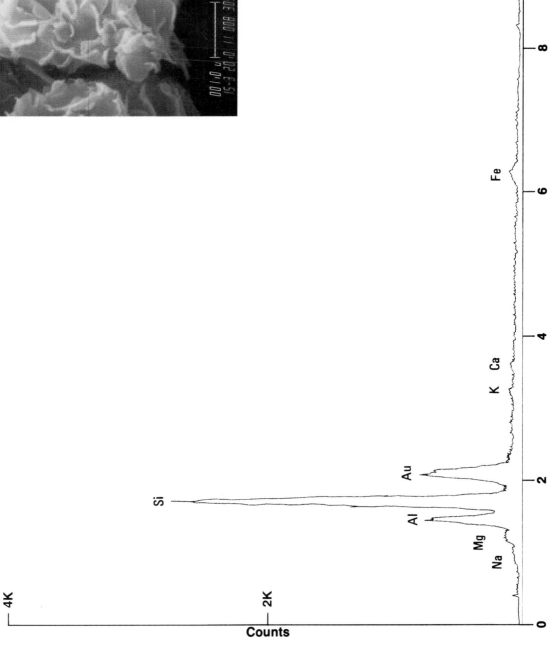

Silicates—Clay (Smectite)

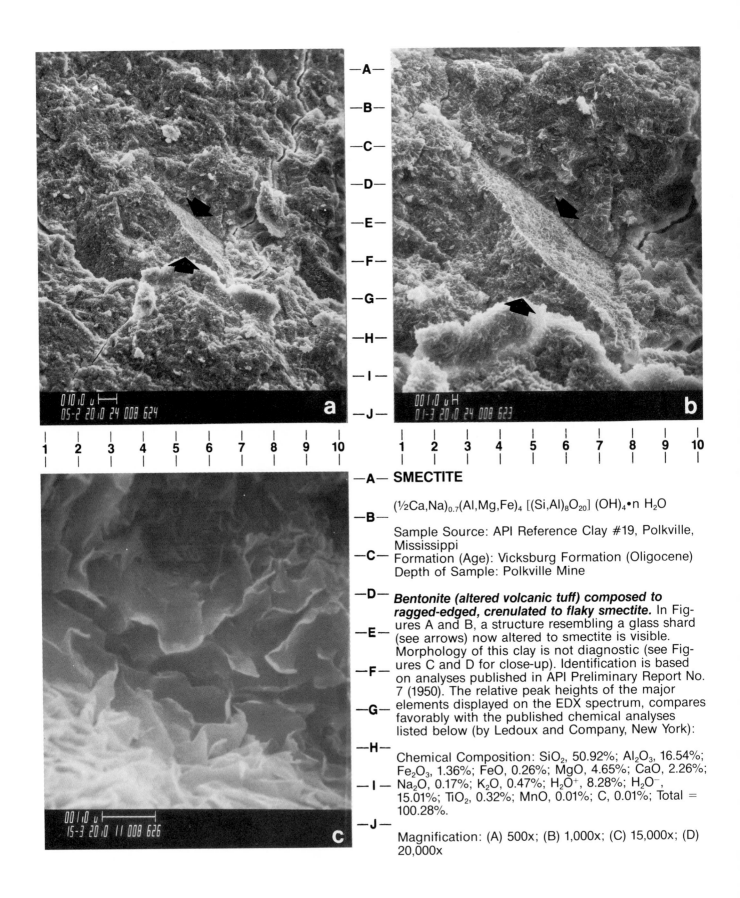

A—
B—
C—
D—
E—
F—
G—
H—
I—
J—

SMECTITE

$(\frac{1}{2}Ca,Na)_{0.7}(Al,Mg,Fe)_4\,[(Si,Al)_8O_{20}]\,(OH)_4 \cdot n\,H_2O$

Sample Source: API Reference Clay #19, Polkville, Mississippi
Formation (Age): Vicksburg Formation (Oligocene)
Depth of Sample: Polkville Mine

Bentonite (altered volcanic tuff) composed to ragged-edged, crenulated to flaky smectite. In Figures A and B, a structure resembling a glass shard (see arrows) now altered to smectite is visible. Morphology of this clay is not diagnostic (see Figures C and D for close-up). Identification is based on analyses published in API Preliminary Report No. 7 (1950). The relative peak heights of the major elements displayed on the EDX spectrum, compares favorably with the published chemical analyses listed below (by Ledoux and Company, New York):

Chemical Composition: SiO_2, 50.92%; Al_2O_3, 16.54%; Fe_2O_3, 1.36%; FeO, 0.26%; MgO, 4.65%; CaO, 2.26%; Na_2O, 0.17%; K_2O, 0.47%; H_2O^+, 8.28%; H_2O^-, 15.01%; TiO_2, 0.32%; MnO, 0.01%; C, 0.01%; Total = 100.28%.

Magnification: (A) 500x; (B) 1,000x; (C) 15,000x; (D) 20,000x

Energy Dispersive X—Ray Spectrum (EDX)

Smectite ($\frac{1}{2}$Ca, Na)$_{0.7}$ (Al, Mg, Fe)$_4$ [(Si, Al)$_8$ O$_{20}$] (OH)$_4$ • n H$_2$O

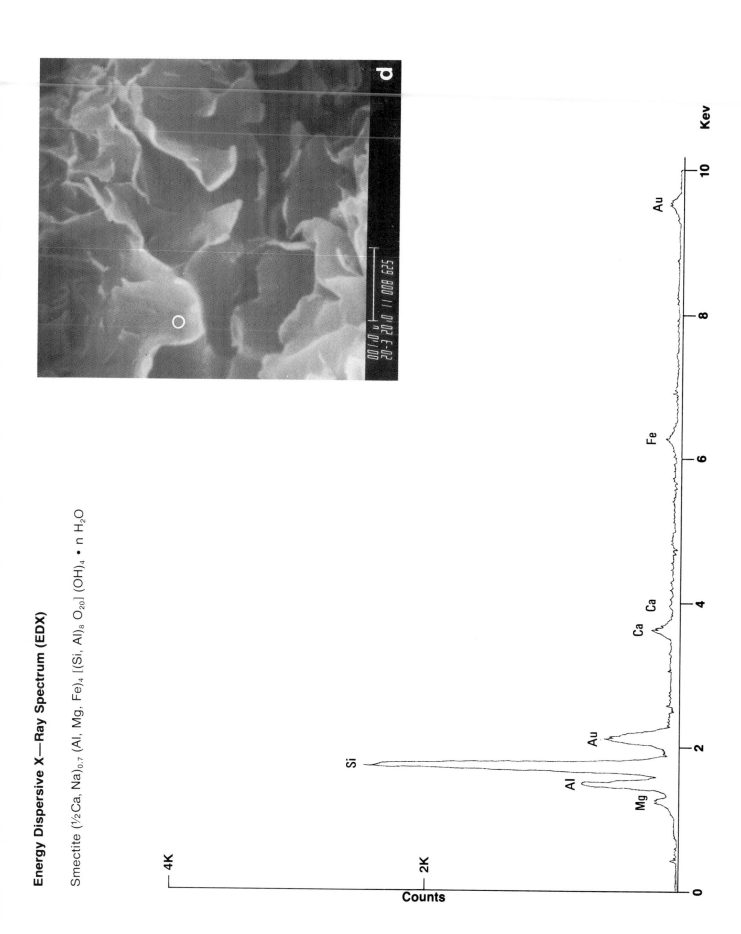

Silicates—Clay (Smectite)

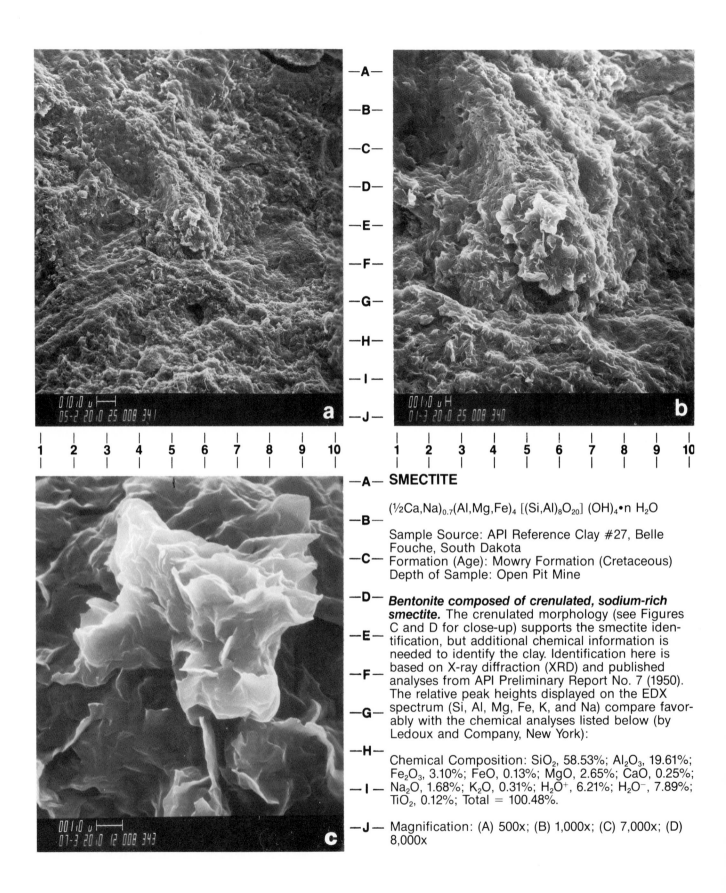

—A— **SMECTITE**

—B— $(\frac{1}{2}Ca,Na)_{0.7}(Al,Mg,Fe)_4 [(Si,Al)_8O_{20}] (OH)_4 \cdot n\ H_2O$

—C— Sample Source: API Reference Clay #27, Belle Fouche, South Dakota
Formation (Age): Mowry Formation (Cretaceous)
Depth of Sample: Open Pit Mine

—D— *Bentonite composed of crenulated, sodium-rich smectite.* The crenulated morphology (see Figures C and D for close-up) supports the smectite iden-
—E— tification, but additional chemical information is needed to identify the clay. Identification here is based on X-ray diffraction (XRD) and published
—F— analyses from API Preliminary Report No. 7 (1950). The relative peak heights displayed on the EDX spectrum (Si, Al, Mg, Fe, K, and Na) compare favor-
—G— ably with the chemical analyses listed below (by Ledoux and Company, New York):

—H— Chemical Composition: SiO_2, 58.53%; Al_2O_3, 19.61%; Fe_2O_3, 3.10%; FeO, 0.13%; MgO, 2.65%; CaO, 0.25%;
—I— Na_2O, 1.68%; K_2O, 0.31%; H_2O^+, 6.21%; H_2O^-, 7.89%; TiO_2, 0.12%; Total = 100.48%.

—J— Magnification: (A) 500x; (B) 1,000x; (C) 7,000x; (D) 8,000x

Energy Dispersive X—Ray Spectrum (EDX)

Smectite ($\frac{1}{2}$Ca, Na)$_{0.7}$ (Al, Mg, Fe)$_4$ [(Si, Al)$_8$ O$_{20}$] (OH)$_4$ · n H$_2$O

Silicates—Clay (Smectite)

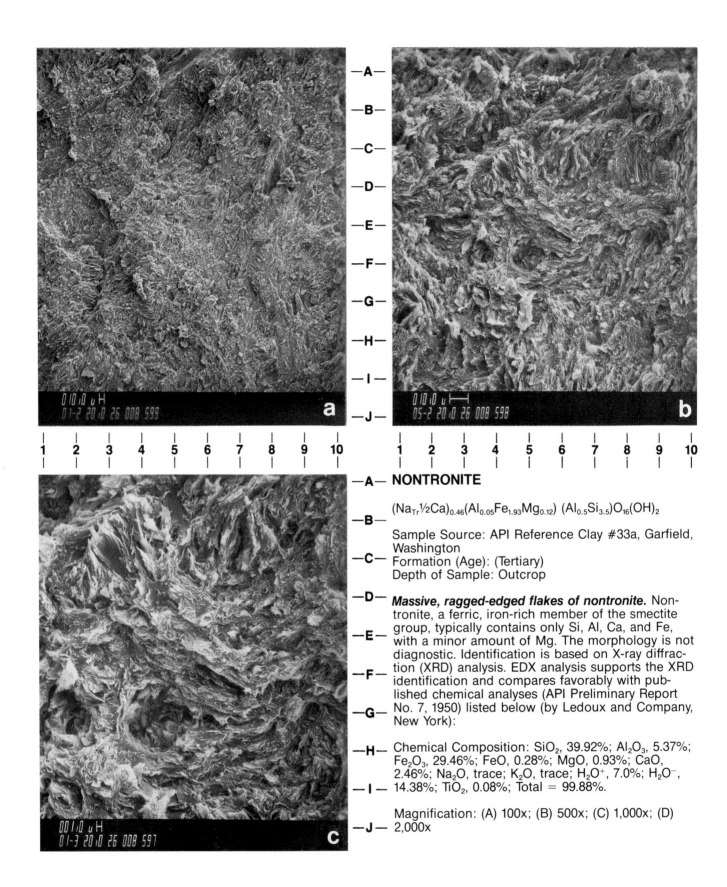

—A— **NONTRONITE**

—B— $(Na_{Tr}\tfrac{1}{2}Ca)_{0.46}(Al_{0.05}Fe_{1.93}Mg_{0.12})\ (Al_{0.5}Si_{3.5})O_{16}(OH)_2$

Sample Source: API Reference Clay #33a, Garfield, Washington

—C— Formation (Age): (Tertiary)
Depth of Sample: Outcrop

—D— *Massive, ragged-edged flakes of nontronite.* Nontronite, a ferric, iron-rich member of the smectite group, typically contains only Si, Al, Ca, and Fe,

—E— with a minor amount of Mg. The morphology is not diagnostic. Identification is based on X-ray diffraction (XRD) analysis. EDX analysis supports the XRD

—F— identification and compares favorably with published chemical analyses (API Preliminary Report No. 7, 1950) listed below (by Ledoux and Company,

—G— New York):

—H— Chemical Composition: SiO_2, 39.92%; Al_2O_3, 5.37%; Fe_2O_3, 29.46%; FeO, 0.28%; MgO, 0.93%; CaO, 2.46%; Na_2O, trace; K_2O, trace; H_2O^+, 7.0%; H_2O^-,

—I— 14.38%; TiO_2, 0.08%; Total = 99.88%.

Magnification: (A) 100x; (B) 500x; (C) 1,000x; (D)

—J— 2,000x

Silicates—Clay (Nontronite)

Energy Dispersive X—Ray Spectrum (EDX)

Nontronite $(Na_{Tr}\frac{1}{2}Ca)_{.46}(Al_{.05}Fe_{1.93}Mg_{.12})(Al_{.50}Si_{3.50})O_{16}(OH)_2$

Silicates—Clay (Nontronite)

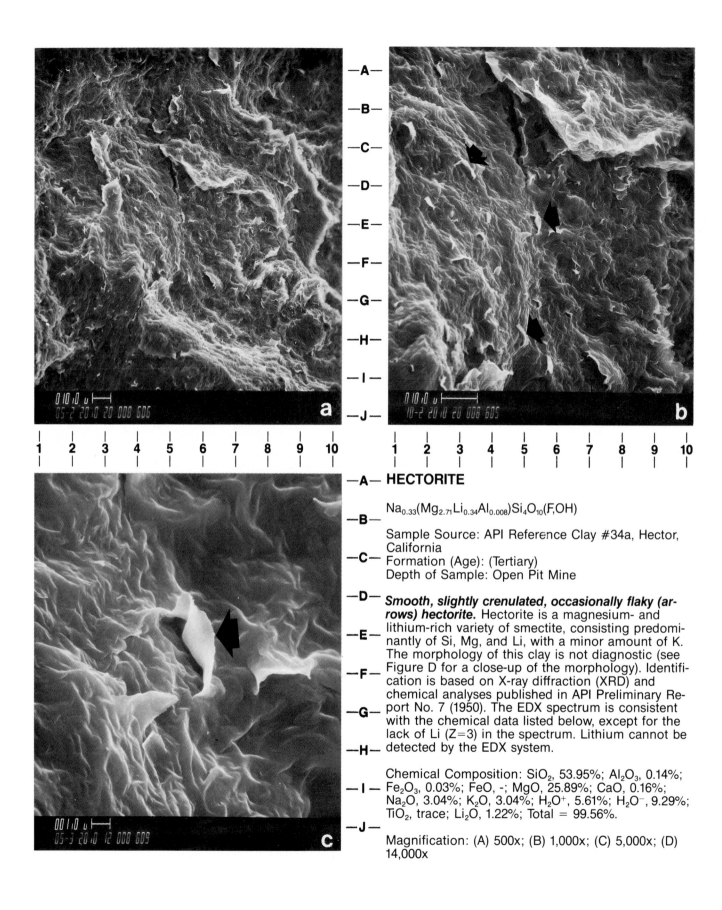

-A- **HECTORITE**

-B- $Na_{0.33}(Mg_{2.71}Li_{0.34}Al_{0.008})Si_4O_{10}(F,OH)$

Sample Source: API Reference Clay #34a, Hector, California
-C- Formation (Age): (Tertiary)
Depth of Sample: Open Pit Mine

-D- ***Smooth, slightly crenulated, occasionally flaky (arrows) hectorite.*** Hectorite is a magnesium- and lithium-rich variety of smectite, consisting predomi-
-E- nantly of Si, Mg, and Li, with a minor amount of K. The morphology of this clay is not diagnostic (see Figure D for a close-up of the morphology). Identifi-
-F- cation is based on X-ray diffraction (XRD) and chemical analyses published in API Preliminary Re-
-G- port No. 7 (1950). The EDX spectrum is consistent with the chemical data listed below, except for the lack of Li (Z=3) in the spectrum. Lithium cannot be
-H- detected by the EDX system.

Chemical Composition: SiO_2, 53.95%; Al_2O_3, 0.14%;
-I- Fe_2O_3, 0.03%; FeO, -; MgO, 25.89%; CaO, 0.16%; Na_2O, 3.04%; K_2O, 3.04%; H_2O^+, 5.61%; H_2O^-, 9.29%; TiO_2, trace; Li_2O, 1.22%; Total = 99.56%.

-J- Magnification: (A) 500x; (B) 1,000x; (C) 5,000x; (D) 14,000x

Energy Dispersive X—Ray Spectrum (EDX)

Hectorite $Na_{0.33}(Mg_{2.71}Li_{.34}Al_{.008})Si_4O_{10}(F, OH)$

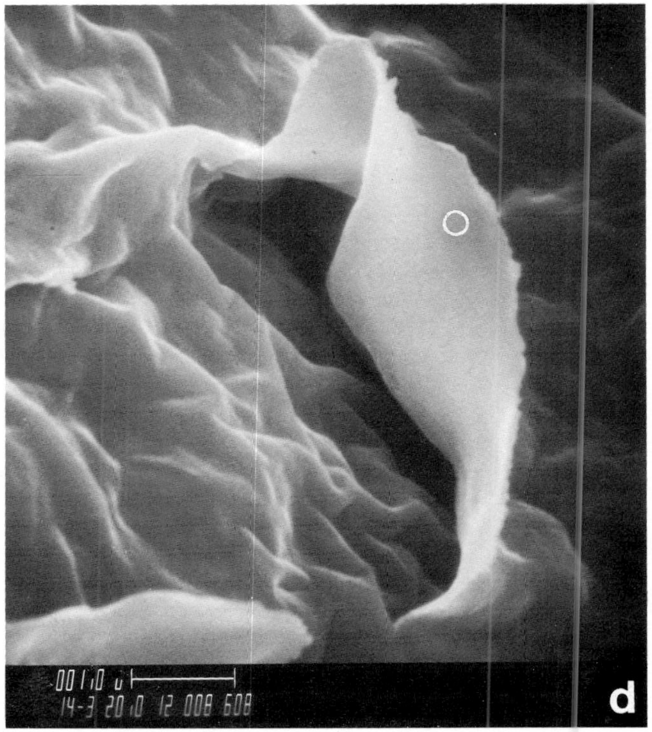

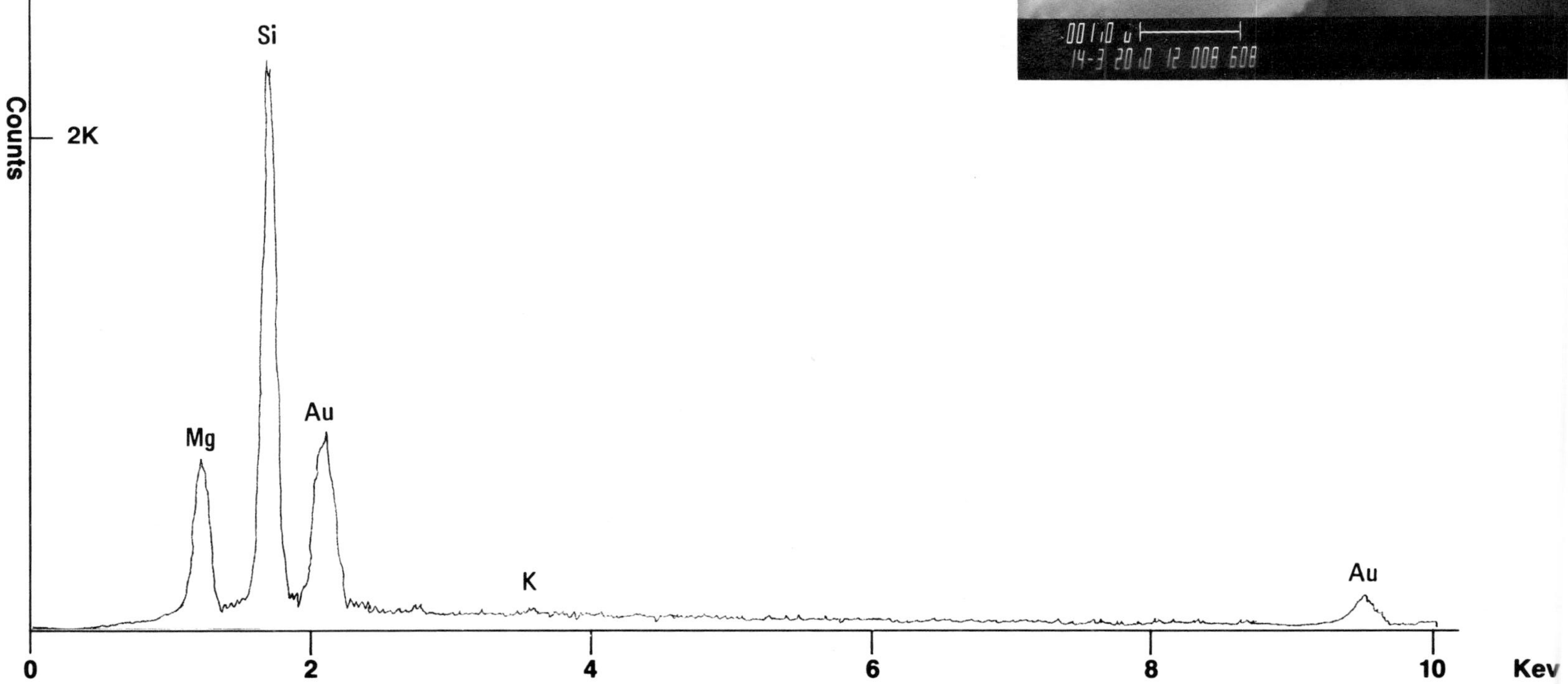

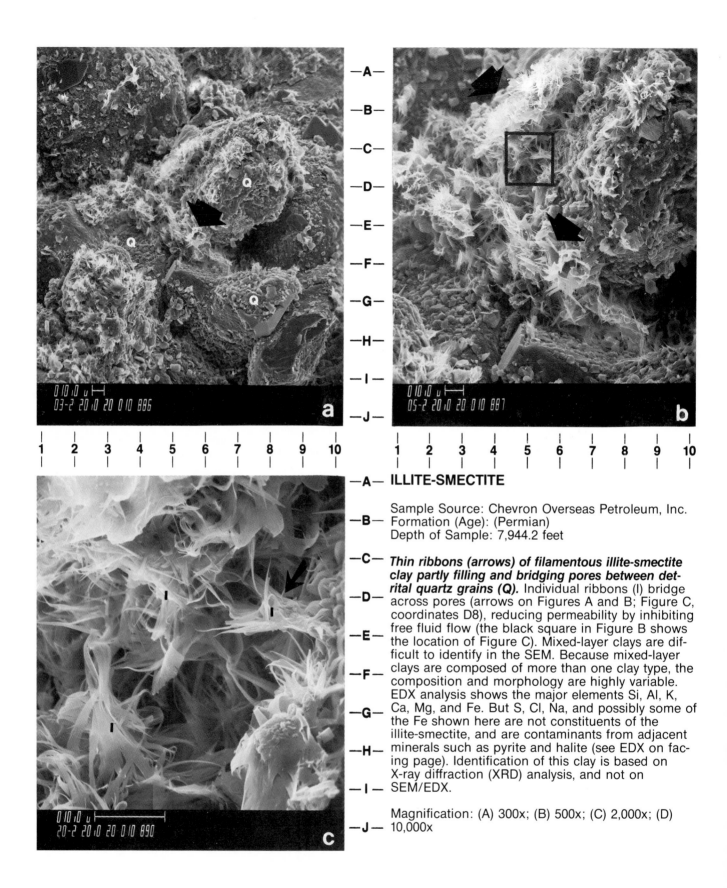

ILLITE-SMECTITE

Sample Source: Chevron Overseas Petroleum, Inc.
Formation (Age): (Permian)
Depth of Sample: 7,944.2 feet

Thin ribbons (arrows) of filamentous illite-smectite clay partly filling and bridging pores between detrital quartz grains (Q). Individual ribbons (I) bridge across pores (arrows on Figures A and B; Figure C, coordinates D8), reducing permeability by inhibiting free fluid flow (the black square in Figure B shows the location of Figure C). Mixed-layer clays are difficult to identify in the SEM. Because mixed-layer clays are composed of more than one clay type, the composition and morphology are highly variable. EDX analysis shows the major elements Si, Al, K, Ca, Mg, and Fe. But S, Cl, Na, and possibly some of the Fe shown here are not constituents of the illite-smectite, and are contaminants from adjacent minerals such as pyrite and halite (see EDX on facing page). Identification of this clay is based on X-ray diffraction (XRD) analysis, and not on SEM/EDX.

Magnification: (A) 300x; (B) 500x; (C) 2,000x; (D) 10,000x

Energy Dispersive X—Ray Spectrum (EDX)

Illite — Smectite

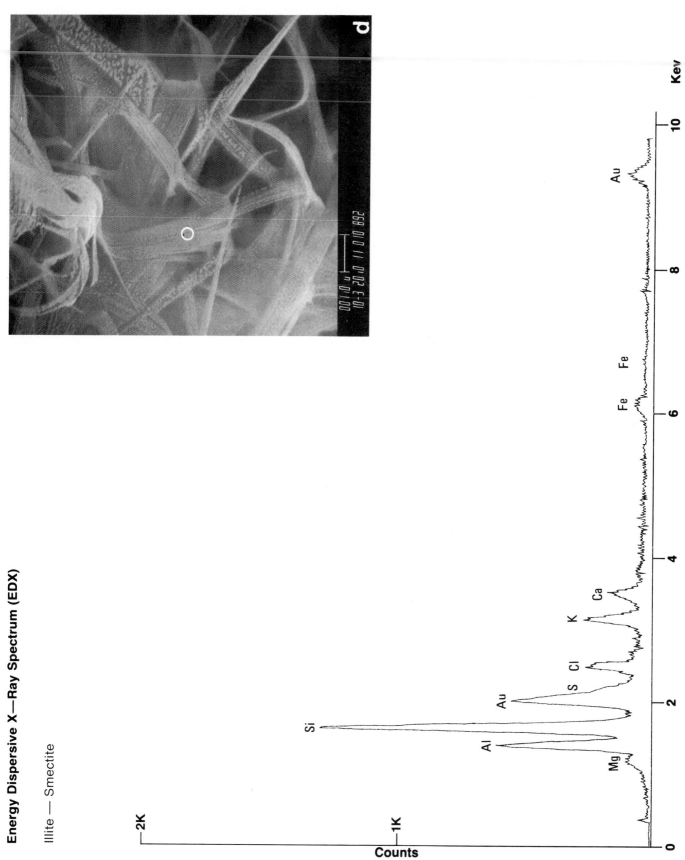

Silicates—Clay (Illite-Smectite)

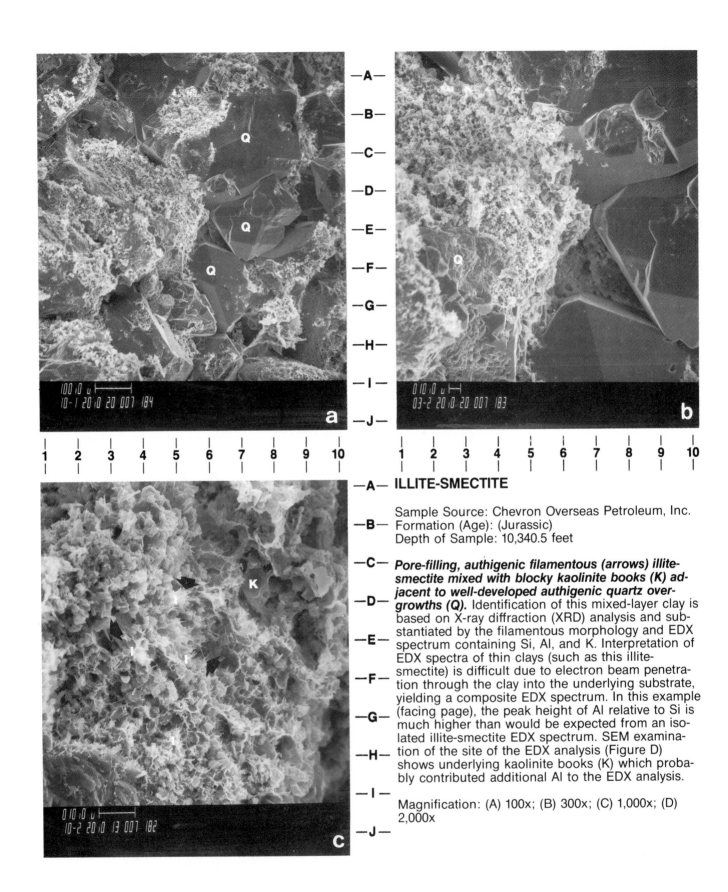

ILLITE-SMECTITE

Sample Source: Chevron Overseas Petroleum, Inc.
Formation (Age): (Jurassic)
Depth of Sample: 10,340.5 feet

Pore-filling, authigenic filamentous (arrows) illite-smectite mixed with blocky kaolinite books (K) adjacent to well-developed authigenic quartz overgrowths (Q). Identification of this mixed-layer clay is based on X-ray diffraction (XRD) analysis and substantiated by the filamentous morphology and EDX spectrum containing Si, Al, and K. Interpretation of EDX spectra of thin clays (such as this illite-smectite) is difficult due to electron beam penetration through the clay into the underlying substrate, yielding a composite EDX spectrum. In this example (facing page), the peak height of Al relative to Si is much higher than would be expected from an isolated illite-smectite EDX spectrum. SEM examination of the site of the EDX analysis (Figure D) shows underlying kaolinite books (K) which probably contributed additional Al to the EDX analysis.

Magnification: (A) 100x; (B) 300x; (C) 1,000x; (D) 2,000x

Energy Dispersive X—Ray Spectrum (EDX)

Illite-Smectite

Silicates—Clay (Illite-Smectite)

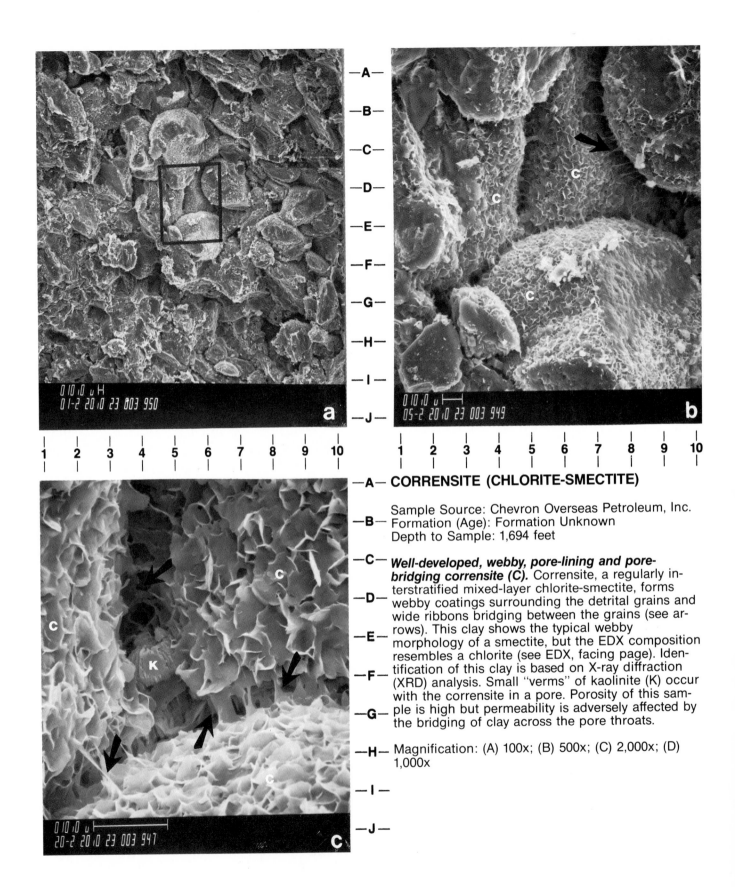

CORRENSITE (CHLORITE-SMECTITE)

Sample Source: Chevron Overseas Petroleum, Inc.
Formation (Age): Formation Unknown
Depth to Sample: 1,694 feet

Well-developed, webby, pore-lining and pore-bridging corrensite (C). Corrensite, a regularly interstratified mixed-layer chlorite-smectite, forms webby coatings surrounding the detrital grains and wide ribbons bridging between the grains (see arrows). This clay shows the typical webby morphology of a smectite, but the EDX composition resembles a chlorite (see EDX, facing page). Identification of this clay is based on X-ray diffraction (XRD) analysis. Small "verms" of kaolinite (K) occur with the corrensite in a pore. Porosity of this sample is high but permeability is adversely affected by the bridging of clay across the pore throats.

Magnification: (A) 100x; (B) 500x; (C) 2,000x; (D) 1,000x

Silicates—Clay (Corrensite)

Energy Dispersive X—Ray Spectrum (EDX)

Corrensite
(Chlorite/Smectite)

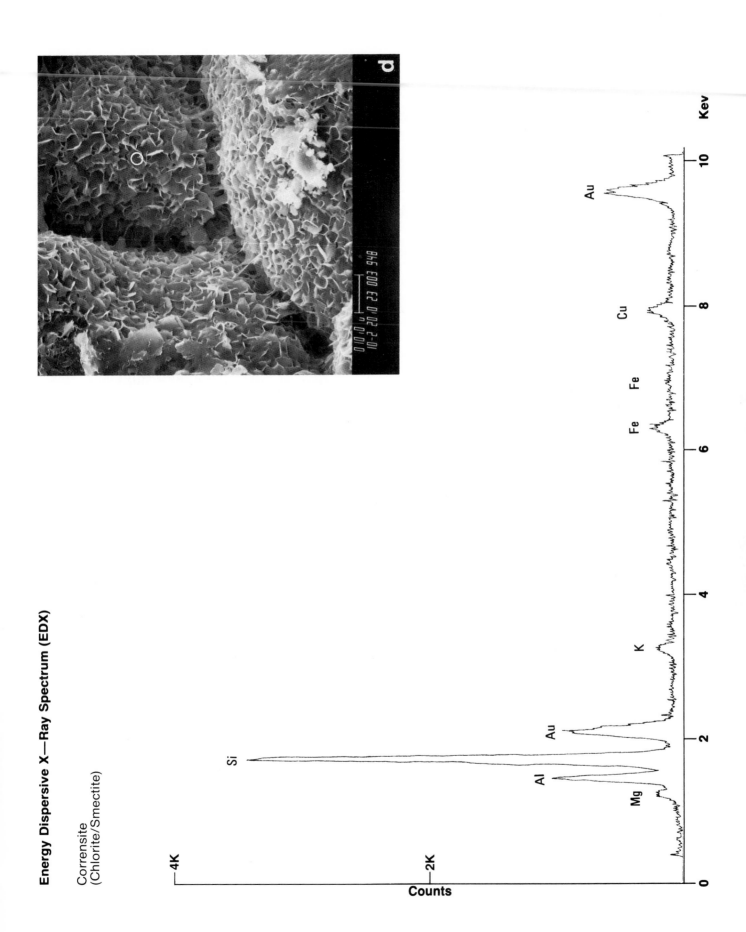

Silicates—Clay (Corrensite)

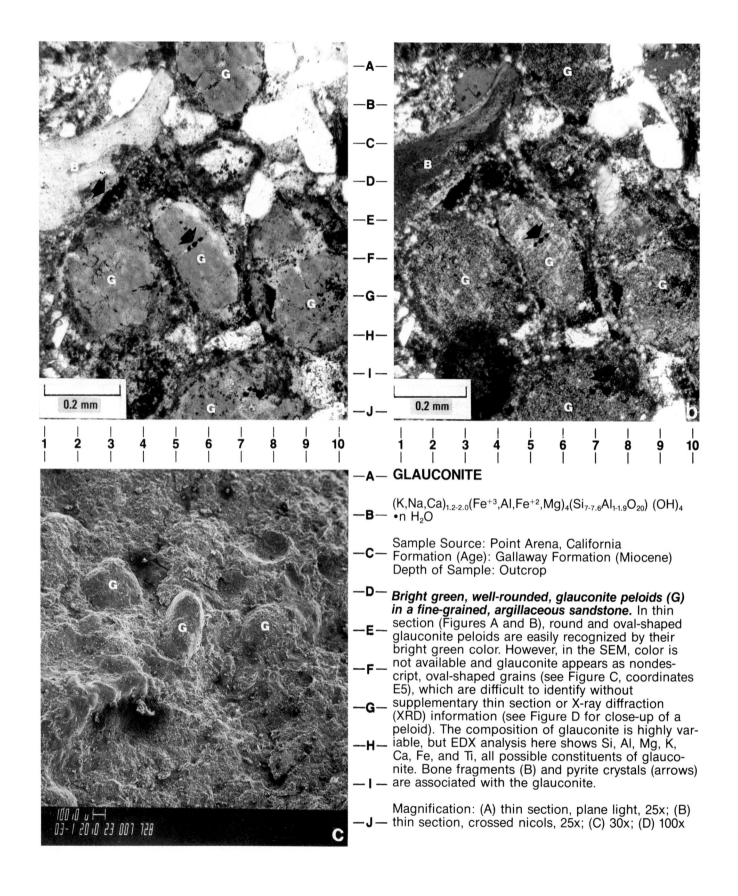

GLAUCONITE

$(K,Na,Ca)_{1.2-2.0}(Fe^{+3},Al,Fe^{+2},Mg)_4(Si_{7-7.6}Al_{1-1.9}O_{20})(OH)_4 \cdot n\,H_2O$

Sample Source: Point Arena, California
Formation (Age): Gallaway Formation (Miocene)
Depth of Sample: Outcrop

Bright green, well-rounded, glauconite peloids (G) in a fine-grained, argillaceous sandstone. In thin section (Figures A and B), round and oval-shaped glauconite peloids are easily recognized by their bright green color. However, in the SEM, color is not available and glauconite appears as nondescript, oval-shaped grains (see Figure C, coordinates E5), which are difficult to identify without supplementary thin section or X-ray diffraction (XRD) information (see Figure D for close-up of a peloid). The composition of glauconite is highly variable, but EDX analysis here shows Si, Al, Mg, K, Ca, Fe, and Ti, all possible constituents of glauconite. Bone fragments (B) and pyrite crystals (arrows) are associated with the glauconite.

Magnification: (A) thin section, plane light, 25x; (B) thin section, crossed nicols, 25x; (C) 30x; (D) 100x

Silicates—Clay (Glauconite)

Energy Dispersive X—Ray Spectrum (EDX)

Glauconite $(K, Na, Ca)_{1.2-2.0} (Fe^{+3}, Al, Fe^{+2}, Mg)_4 (Si_{7-7.6} Al_{1-1.9} O_{20}) (OH)_4 \cdot n\ H_2O$

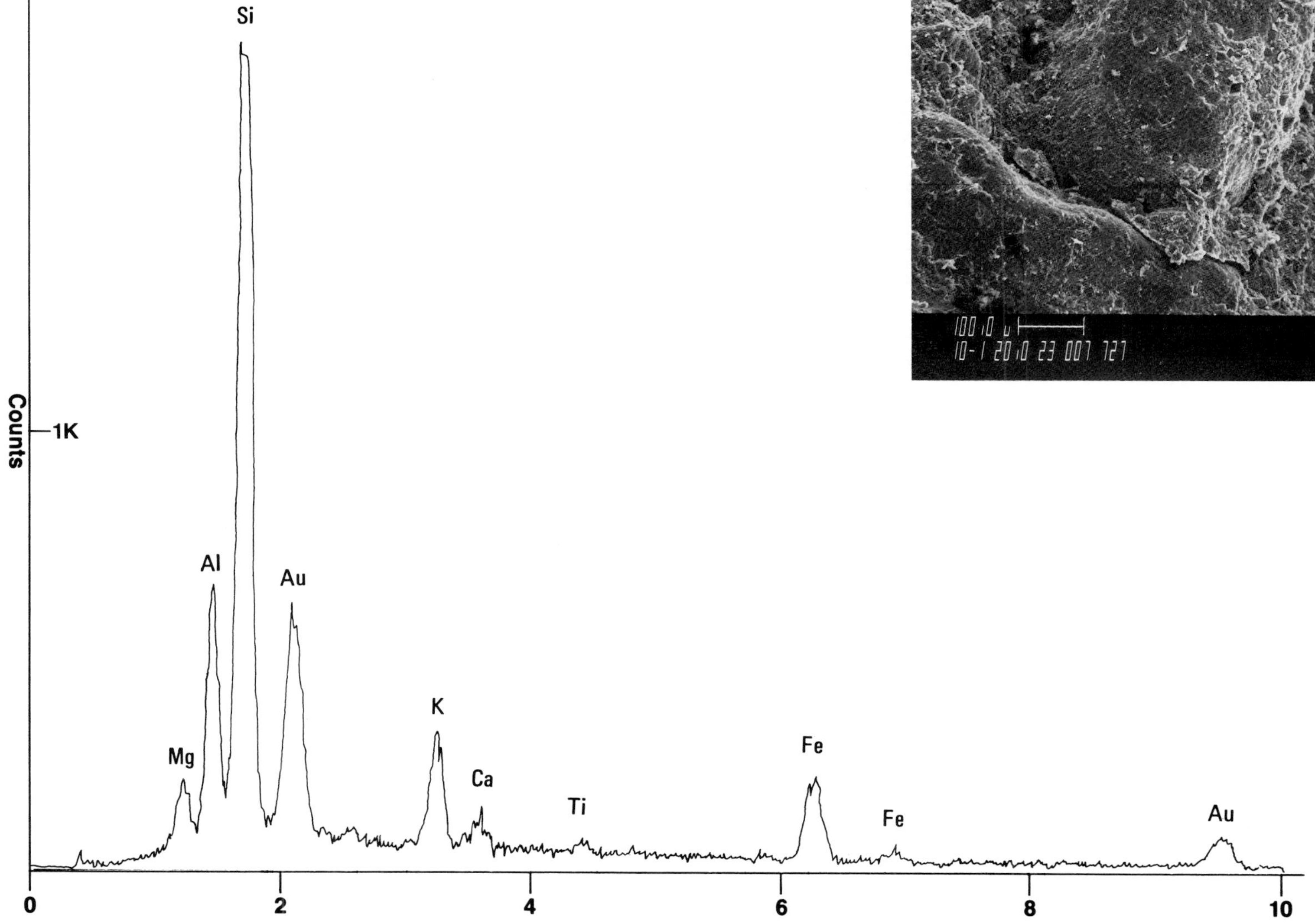

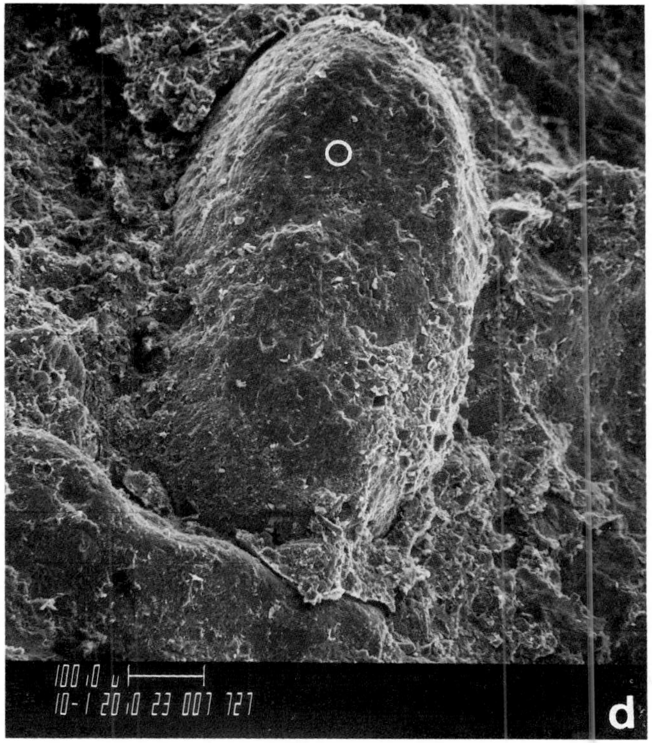

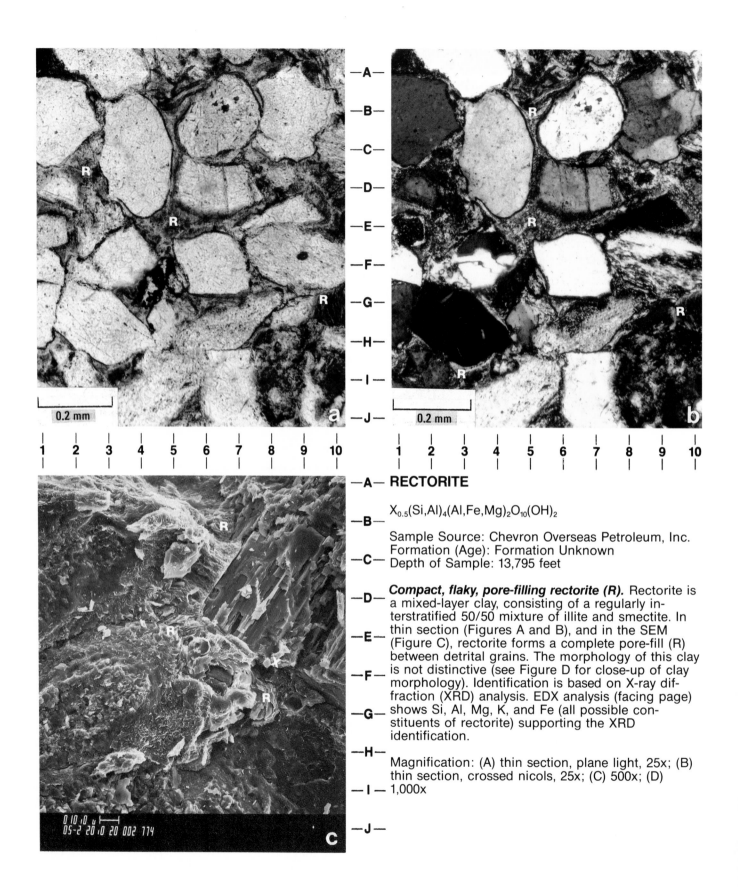

—A— **RECTORITE**

—B— $X_{0.5}(Si,Al)_4(Al,Fe,Mg)_2O_{10}(OH)_2$

Sample Source: Chevron Overseas Petroleum, Inc.
Formation (Age): Formation Unknown
—C— Depth of Sample: 13,795 feet

—D— ***Compact, flaky, pore-filling rectorite (R).*** Rectorite is a mixed-layer clay, consisting of a regularly interstratified 50/50 mixture of illite and smectite. In thin section (Figures A and B), and in the SEM (Figure C), rectorite forms a complete pore-fill (R) between detrital grains. The morphology of this clay is not distinctive (see Figure D for close-up of clay morphology). Identification is based on X-ray diffraction (XRD) analysis. EDX analysis (facing page) shows Si, Al, Mg, K, and Fe (all possible constituents of rectorite) supporting the XRD identification.

—H— Magnification: (A) thin section, plane light, 25x; (B) thin section, crossed nicols, 25x; (C) 500x; (D) —I— 1,000x

—J—

Energy Dispersive X—Ray Spectrum (EDX)

Rectorite $X_{0.5}$ (Si, Al)$_4$ (Al, Fe, Mg)$_2$ O$_{10}$ (OH)$_2$

Silicates—Clay (Rectorite)

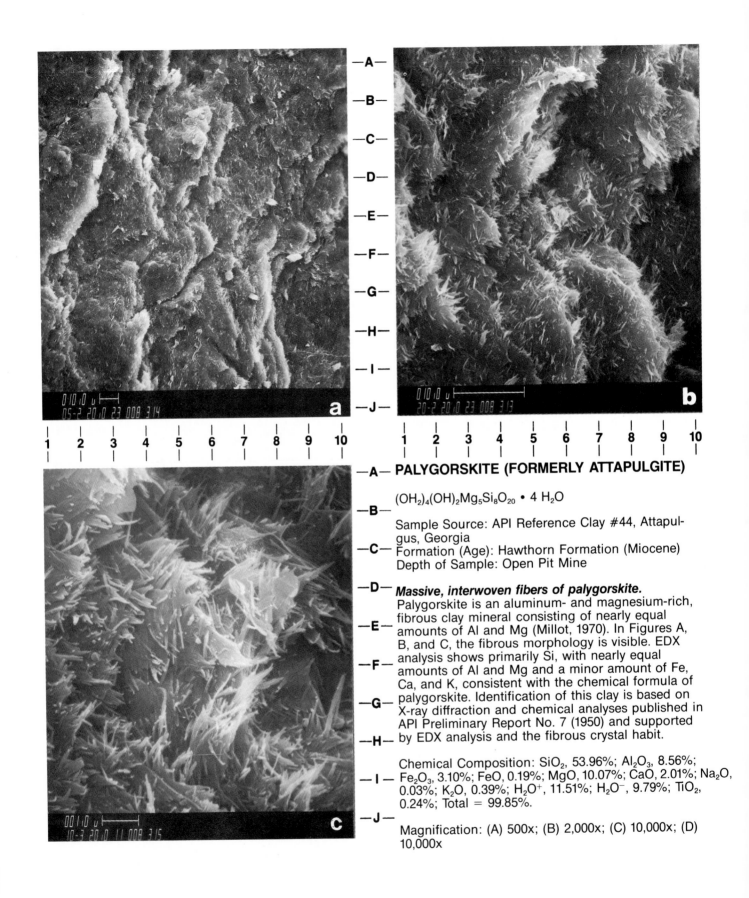

PALYGORSKITE (FORMERLY ATTAPULGITE)

$(OH_2)_4(OH)_2Mg_5Si_8O_{20} \cdot 4\ H_2O$

Sample Source: API Reference Clay #44, Attapulgus, Georgia
Formation (Age): Hawthorn Formation (Miocene)
Depth of Sample: Open Pit Mine

Massive, interwoven fibers of palygorskite.
Palygorskite is an aluminum- and magnesium-rich, fibrous clay mineral consisting of nearly equal amounts of Al and Mg (Millot, 1970). In Figures A, B, and C, the fibrous morphology is visible. EDX analysis shows primarily Si, with nearly equal amounts of Al and Mg and a minor amount of Fe, Ca, and K, consistent with the chemical formula of palygorskite. Identification of this clay is based on X-ray diffraction and chemical analyses published in API Preliminary Report No. 7 (1950) and supported by EDX analysis and the fibrous crystal habit.

Chemical Composition: SiO_2, 53.96%; Al_2O_3, 8.56%; Fe_2O_3, 3.10%; FeO, 0.19%; MgO, 10.07%; CaO, 2.01%; Na_2O, 0.03%; K_2O, 0.39%; H_2O^+, 11.51%; H_2O^-, 9.79%; TiO_2, 0.24%; Total = 99.85%.

Magnification: (A) 500x; (B) 2,000x; (C) 10,000x; (D) 10,000x

Silicates—Clay (Palygorskite)

Energy Dispersive X—Ray Spectrum (EDX)

Palygorskite $(OH_2)_4 (OH)_2 Mg_5 Si_8 O_{20} \cdot 4 H_2O$

Energy Dispersive X—Ray Spectrum (EDX)

Vermiculite $(Mg, Ca)_{0.7} (Mg, Fe^{+3}, Al)_6 [(Al, Si)_8 O_{20}] (OH)_4 \cdot 8 H_2O$

Silicates—Clay (Vermiculite)

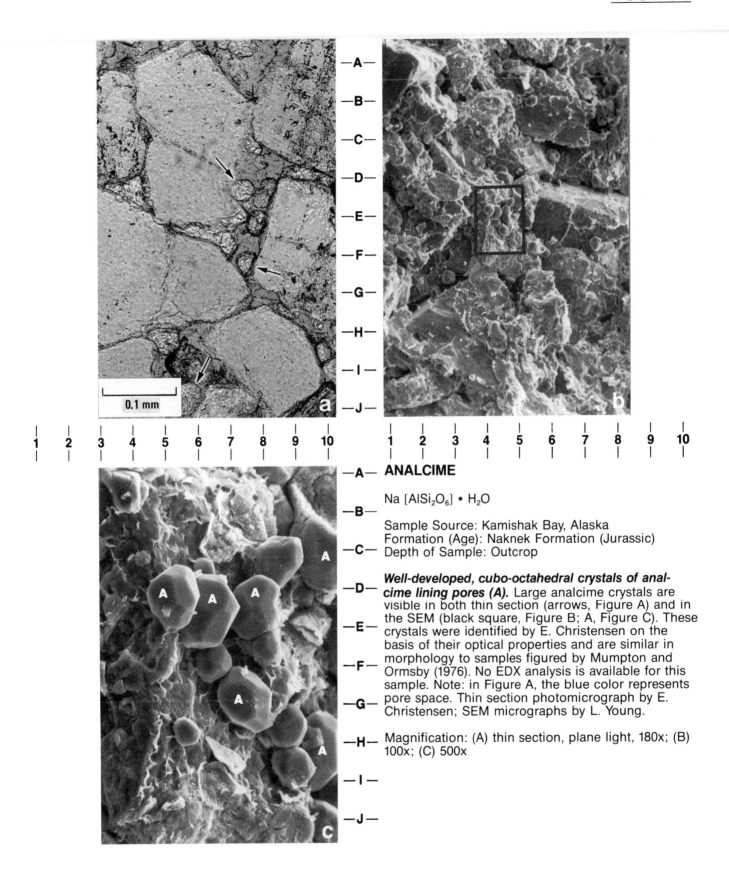

ANALCIME

Na [AlSi₂O₆] • H₂O

$$Na\,[AlSi_2O_6] \cdot H_2O$$

Sample Source: Kamishak Bay, Alaska
Formation (Age): Naknek Formation (Jurassic)
Depth of Sample: Outcrop

Well-developed, cubo-octahedral crystals of analcime lining pores (A). Large analcime crystals are visible in both thin section (arrows, Figure A) and in the SEM (black square, Figure B; A, Figure C). These crystals were identified by E. Christensen on the basis of their optical properties and are similar in morphology to samples figured by Mumpton and Ormsby (1976). No EDX analysis is available for this sample. Note: in Figure A, the blue color represents pore space. Thin section photomicrograph by E. Christensen; SEM micrographs by L. Young.

Magnification: (A) thin section, plane light, 180x; (B) 100x; (C) 500x

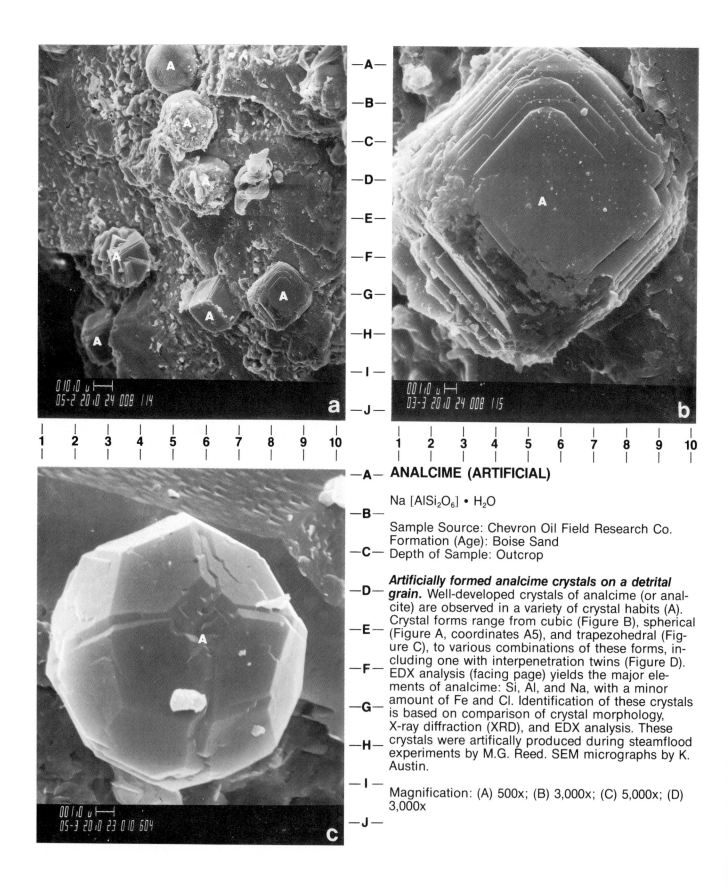

ANALCIME (ARTIFICIAL)

Na [AlSi$_2$O$_6$] • H$_2$O

Sample Source: Chevron Oil Field Research Co.
Formation (Age): Boise Sand
Depth of Sample: Outcrop

Artificially formed analcime crystals on a detrital grain. Well-developed crystals of analcime (or analcite) are observed in a variety of crystal habits (A). Crystal forms range from cubic (Figure B), spherical (Figure A, coordinates A5), and trapezohedral (Figure C), to various combinations of these forms, including one with interpenetration twins (Figure D). EDX analysis (facing page) yields the major elements of analcime: Si, Al, and Na, with a minor amount of Fe and Cl. Identification of these crystals is based on comparison of crystal morphology, X-ray diffraction (XRD), and EDX analysis. These crystals were artifically produced during steamflood experiments by M.G. Reed. SEM micrographs by K. Austin.

Magnification: (A) 500x; (B) 3,000x; (C) 5,000x; (D) 3,000x

Silicates—Zeolite (Analcime)

Energy Dispersive X—Ray Spectrum (EDX)

Analcime Na [Al Si$_2$ O$_6$] • H$_2$O

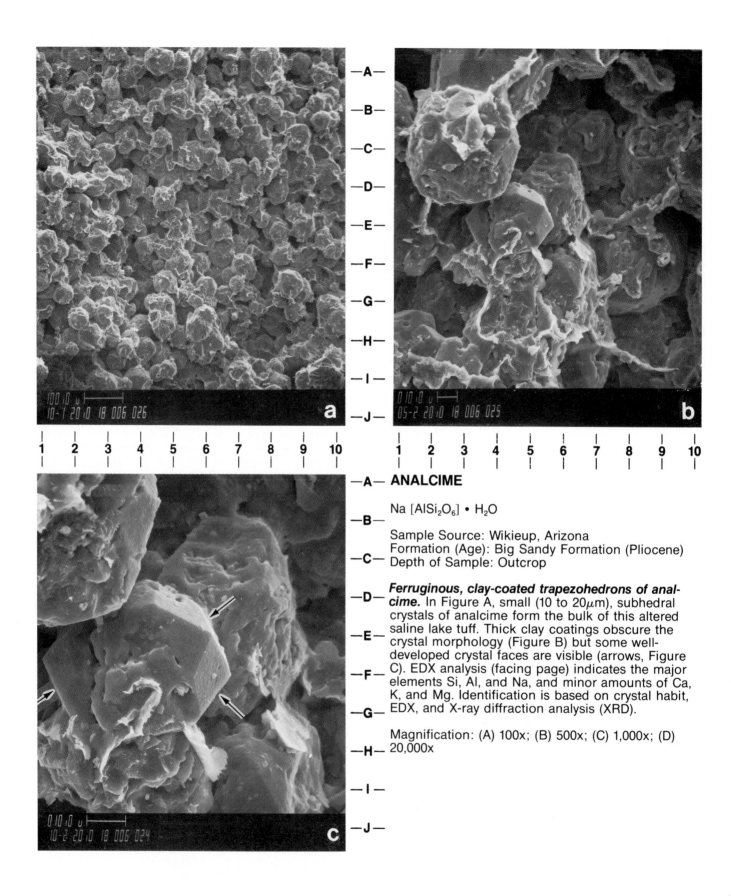

ANALCIME

Na [AlSi$_2$O$_6$] • H$_2$O

Sample Source: Wikieup, Arizona
Formation (Age): Big Sandy Formation (Pliocene)
Depth of Sample: Outcrop

Ferruginous, clay-coated trapezohedrons of analcime. In Figure A, small (10 to 20μm), subhedral crystals of analcime form the bulk of this altered saline lake tuff. Thick clay coatings obscure the crystal morphology (Figure B) but some well-developed crystal faces are visible (arrows, Figure C). EDX analysis (facing page) indicates the major elements Si, Al, and Na, and minor amounts of Ca, K, and Mg. Identification is based on crystal habit, EDX, and X-ray diffraction analysis (XRD).

Magnification: (A) 100x; (B) 500x; (C) 1,000x; (D) 20,000x

Energy Dispersive X—Ray Spectrum (EDX)

Analcime Na [Al Si$_2$ O$_6$] • H$_2$O

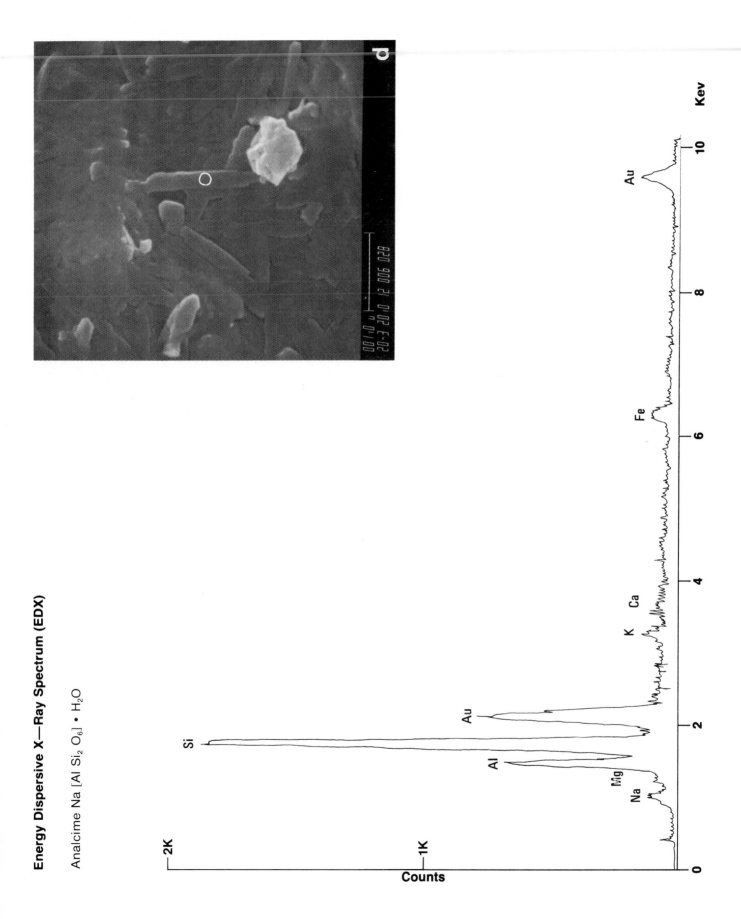

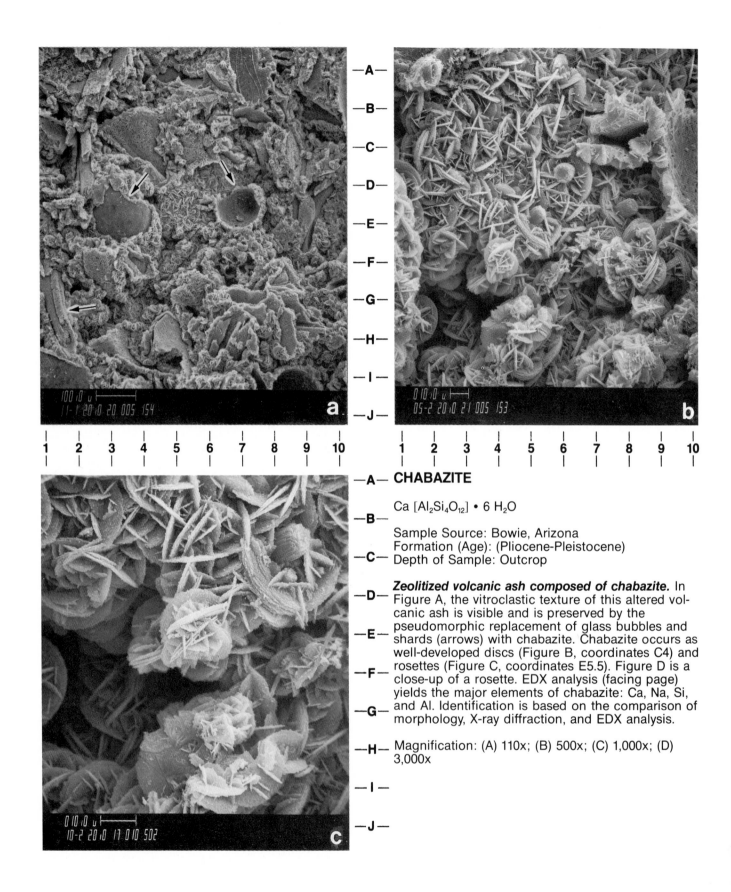

A
B
C
D
E
F
G
H
I
J

1 2 3 4 5 6 7 8 9 10

1 2 3 4 5 6 7 8 9 10

A— **CHABAZITE**

B— Ca [Al$_2$Si$_4$O$_{12}$] • 6 H$_2$O

C— Sample Source: Bowie, Arizona
Formation (Age): (Pliocene-Pleistocene)
Depth of Sample: Outcrop

D— ***Zeolitized volcanic ash composed of chabazite.*** In Figure A, the vitroclastic texture of this altered volcanic ash is visible and is preserved by the pseudomorphic replacement of glass bubbles and shards (arrows) with chabazite. Chabazite occurs as well-developed discs (Figure B, coordinates C4) and rosettes (Figure C, coordinates E5.5). Figure D is a close-up of a rosette. EDX analysis (facing page) yields the major elements of chabazite: Ca, Na, Si, and Al. Identification is based on the comparison of morphology, X-ray diffraction, and EDX analysis.

H— Magnification: (A) 110x; (B) 500x; (C) 1,000x; (D) 3,000x

I
J

Energy Dispersive X—Ray Spectrum (EDX)

Chabazite Ca $[Al_2 Si_4 O_{12}] \cdot 6 H_2O$

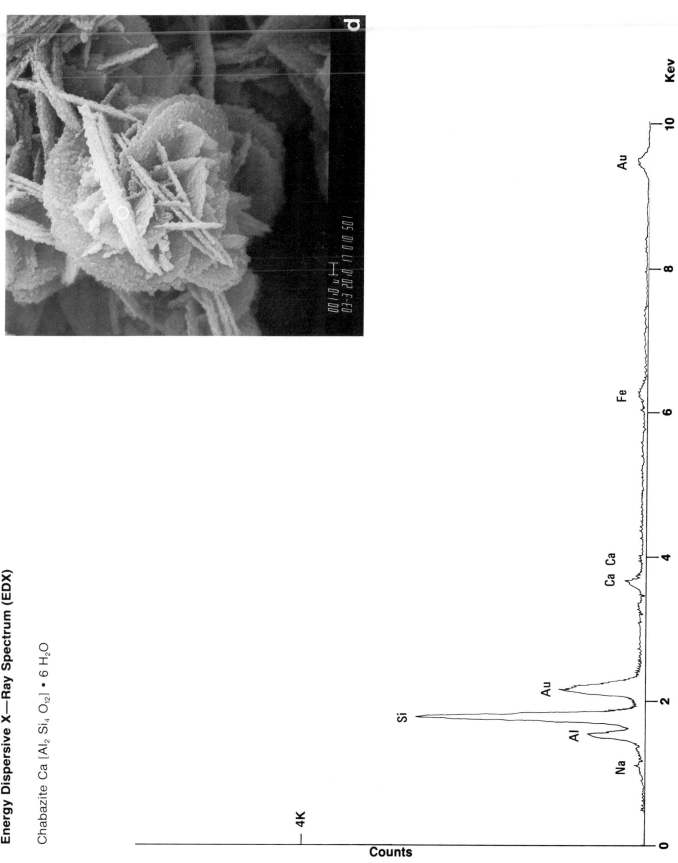

Silicates—Zeolite (Chabazite)

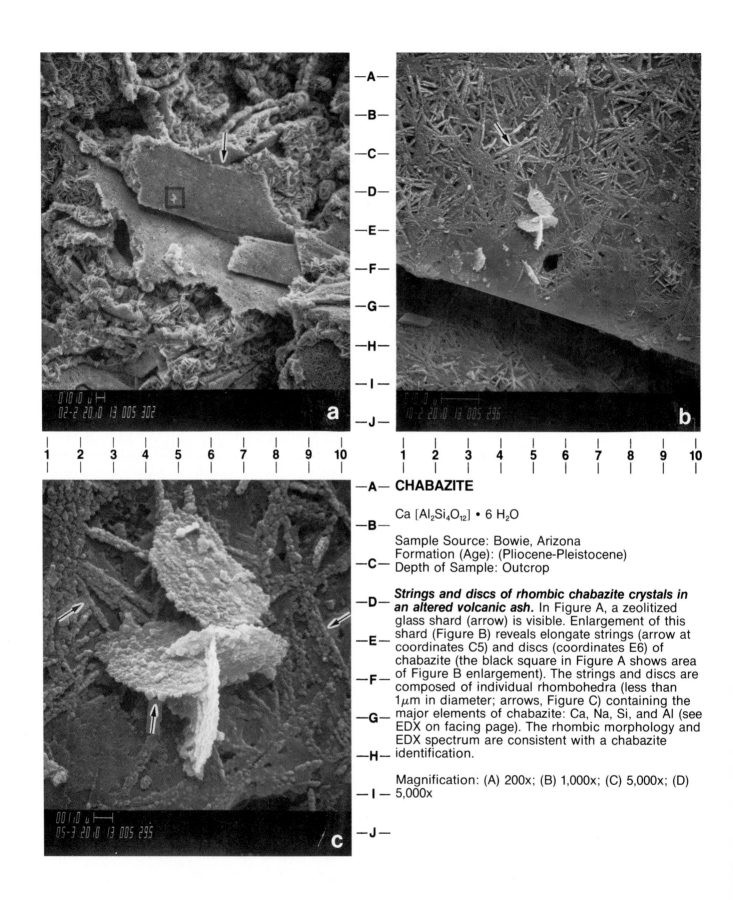

| 1 | 2 | 3 | 4 | 5 | 6 | 7 | 8 | 9 | 10 |

—A— **CHABAZITE**

—B— Ca [Al₂Si₄O₁₂] • 6 H₂O

Ca $[Al_2Si_4O_{12}]$ • 6 H_2O

Sample Source: Bowie, Arizona
Formation (Age): (Pliocene-Pleistocene)
—C— Depth of Sample: Outcrop

—D— ***Strings and discs of rhombic chabazite crystals in an altered volcanic ash.*** In Figure A, a zeolitized glass shard (arrow) is visible. Enlargement of this —E— shard (Figure B) reveals elongate strings (arrow at coordinates C5) and discs (coordinates E6) of chabazite (the black square in Figure A shows area —F— of Figure B enlargement). The strings and discs are composed of individual rhombohedra (less than 1μm in diameter; arrows, Figure C) containing the —G— major elements of chabazite: Ca, Na, Si, and Al (see EDX on facing page). The rhombic morphology and EDX spectrum are consistent with a chabazite —H— identification.

Magnification: (A) 200x; (B) 1,000x; (C) 5,000x; (D) —I— 5,000x

—J—

108

Silicates—Zeolite (Chabazite)

Energy Dispersive X—Ray Spectrum (EDX)

Chabazite Ca [Al$_2$ Si$_4$ O$_{12}$] • 6 H$_2$O

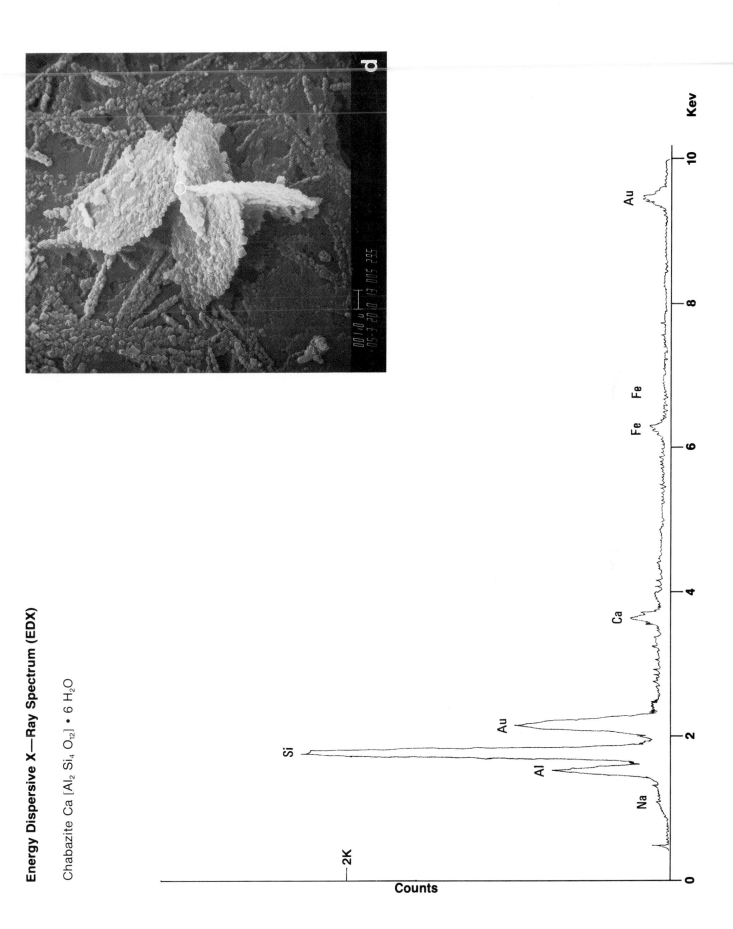

Silicates—Zeolite (Chabazite)

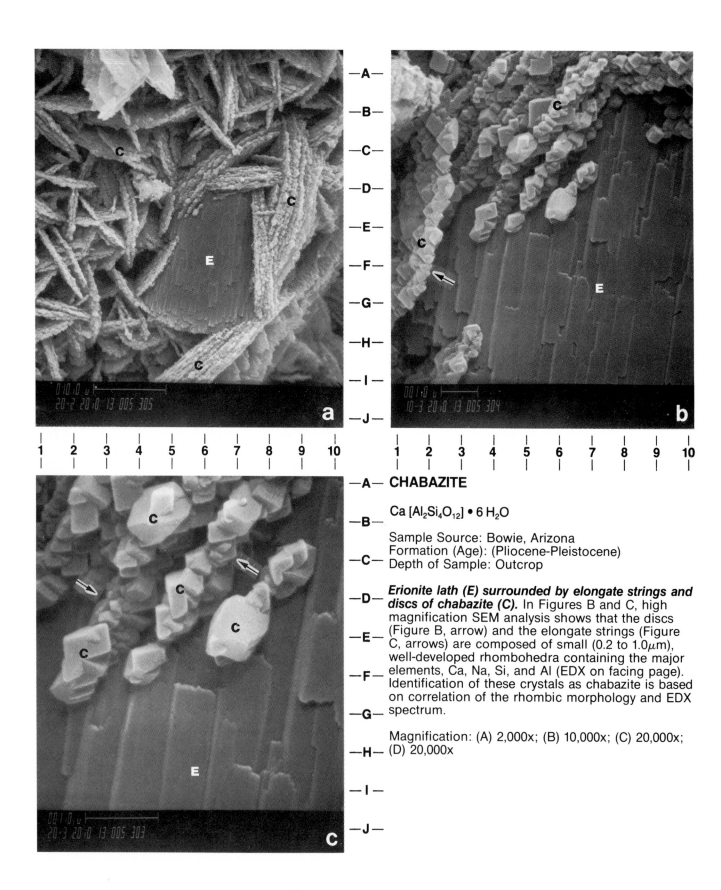

CHABAZITE

Ca [Al$_2$Si$_4$O$_{12}$] • 6 H$_2$O

Sample Source: Bowie, Arizona
Formation (Age): (Pliocene-Pleistocene)
Depth of Sample: Outcrop

Erionite lath (E) surrounded by elongate strings and discs of chabazite (C). In Figures B and C, high magnification SEM analysis shows that the discs (Figure B, arrow) and the elongate strings (Figure C, arrows) are composed of small (0.2 to 1.0μm), well-developed rhombohedra containing the major elements, Ca, Na, Si, and Al (EDX on facing page). Identification of these crystals as chabazite is based on correlation of the rhombic morphology and EDX spectrum.

Magnification: (A) 2,000x; (B) 10,000x; (C) 20,000x; (D) 20,000x

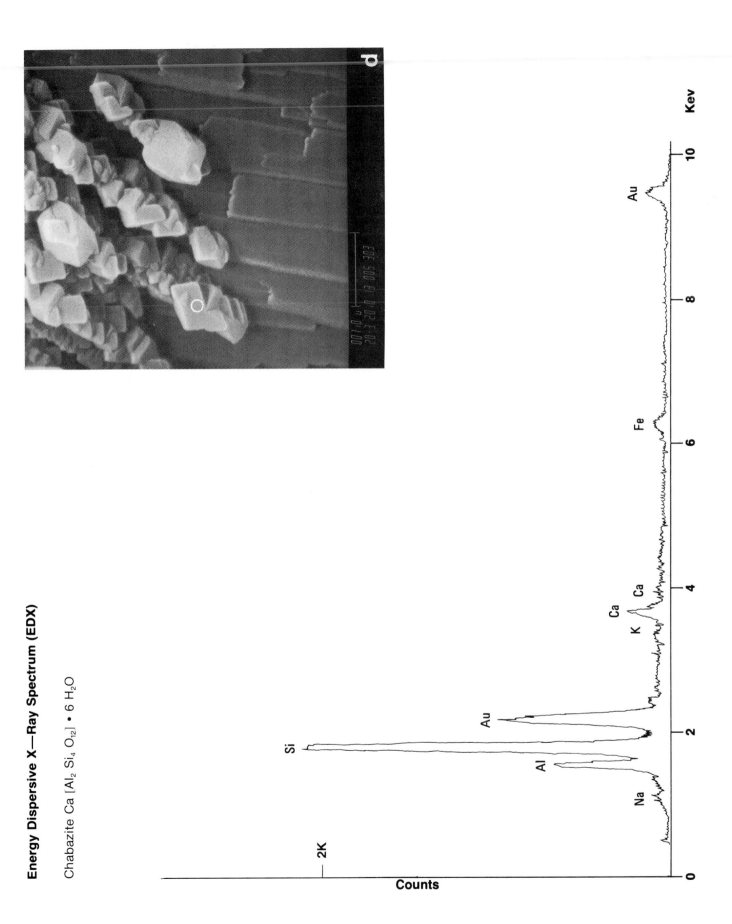

Energy Dispersive X—Ray Spectrum (EDX)

Chabazite Ca [Al$_2$ Si$_4$ O$_{12}$] • 6 H$_2$O

Silicates—Zeolite (Chabazite)

111

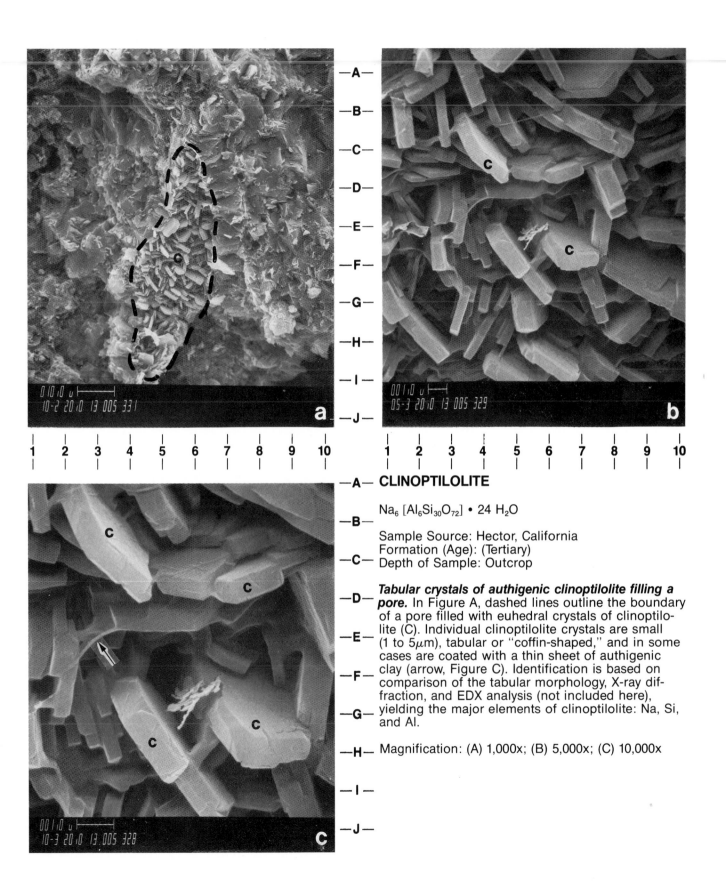

CLINOPTILOLITE

Na$_6$ [Al$_6$Si$_{30}$O$_{72}$] • 24 H$_2$O

Sample Source: Hector, California
Formation (Age): (Tertiary)
Depth of Sample: Outcrop

Tabular crystals of authigenic clinoptilolite filling a pore. In Figure A, dashed lines outline the boundary of a pore filled with euhedral crystals of clinoptilolite (C). Individual clinoptilolite crystals are small (1 to 5μm), tabular or "coffin-shaped," and in some cases are coated with a thin sheet of authigenic clay (arrow, Figure C). Identification is based on comparison of the tabular morphology, X-ray diffraction, and EDX analysis (not included here), yielding the major elements of clinoptilolite: Na, Si, and Al.

Magnification: (A) 1,000x; (B) 5,000x; (C) 10,000x

Silicates—Zeolite (Clinoptilolite)

CLINOPTILOLITE

Na$_6$ [Al$_6$Si$_{30}$O$_{72}$] • 24 H$_2$O

Sample Source: Castle Creek, Idaho
Formation (Age): Chalk Hills (Miocene-Pliocene)
Depth of Sample: Outcrop

Authigenic clinoptilolite lining pores in a lacustrine tuff. In Figures A, B, and C, small (1 to 10μm), tabular, euhedral crystals of clinoptilolite (C) are shown lining a pore. EDX analysis (facing page) yields the major elements of clinoptilolite: Na, K, Si, and Al. Correlation of the tabular morphology, X-ray diffraction analysis, and EDX analysis was used to identify these crystals as clinoptilolite.

Magnification: (A) 500x; (B) 1,000x; (C) 10,000x; (D) 20,000x

Energy Dispersive X—Ray Spectrum (EDX)

Clinoptilolite $Na_6 [Al_6 Si_{30} O_{72}] \cdot 24 H_2O$

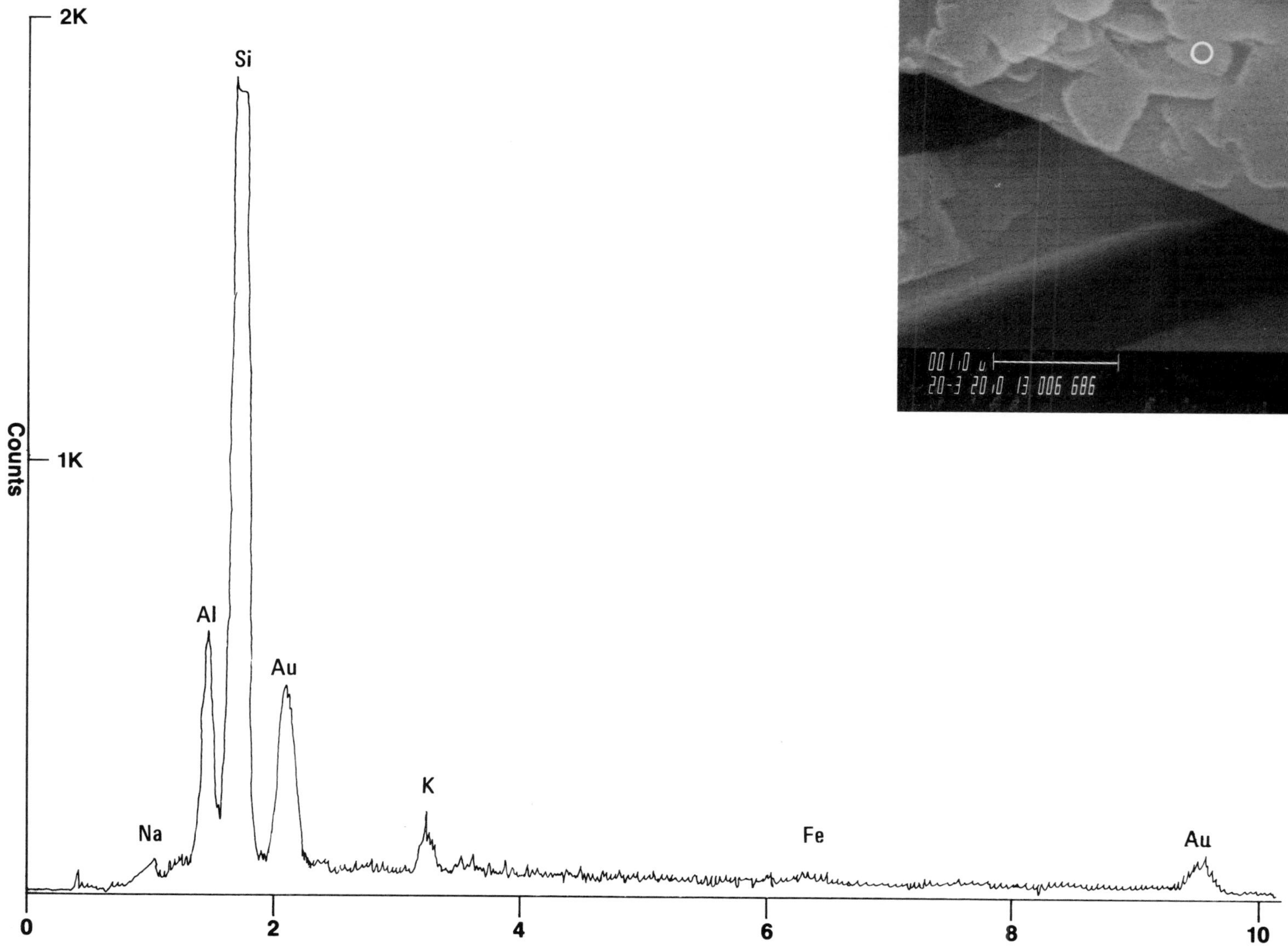

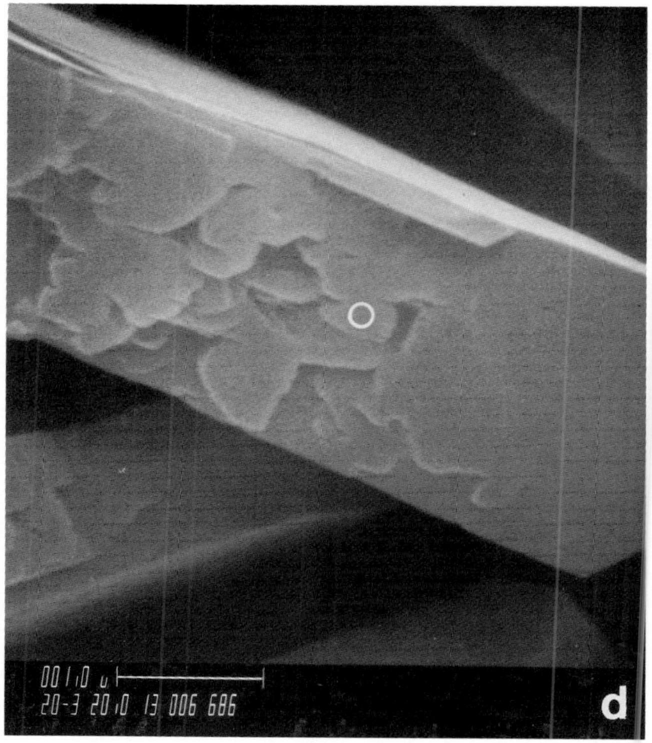

Silicates—Zeolite (Clinoptilolite)

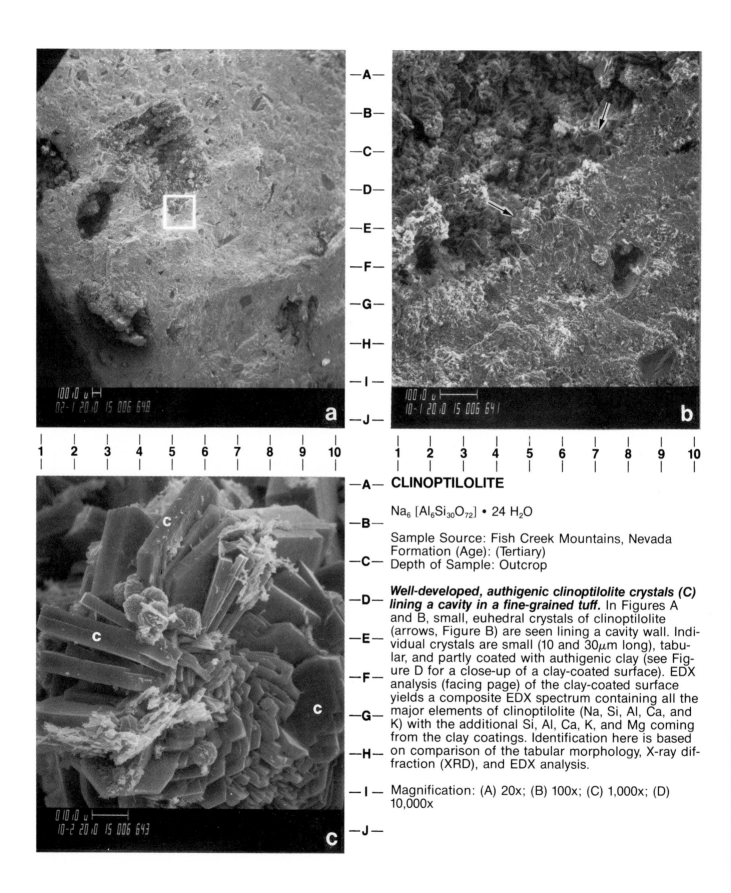

CLINOPTILOLITE

Na$_6$ [Al$_6$Si$_{30}$O$_{72}$] • 24 H$_2$O

Sample Source: Fish Creek Mountains, Nevada
Formation (Age): (Tertiary)
Depth of Sample: Outcrop

Well-developed, authigenic clinoptilolite crystals (C) lining a cavity in a fine-grained tuff. In Figures A and B, small, euhedral crystals of clinoptilolite (arrows, Figure B) are seen lining a cavity wall. Individual crystals are small (10 and 30μm long), tabular, and partly coated with authigenic clay (see Figure D for a close-up of a clay-coated surface). EDX analysis (facing page) of the clay-coated surface yields a composite EDX spectrum containing all the major elements of clinoptilolite (Na, Si, Al, Ca, and K) with the additional Si, Al, Ca, K, and Mg coming from the clay coatings. Identification here is based on comparison of the tabular morphology, X-ray diffraction (XRD), and EDX analysis.

Magnification: (A) 20x; (B) 100x; (C) 1,000x; (D) 10,000x

Energy Dispersive X—Ray Spectrum (EDX)

Clinoptilolite $Na_6 [Al_6 Si_{30} O_{72}] \cdot 24 H_2O$

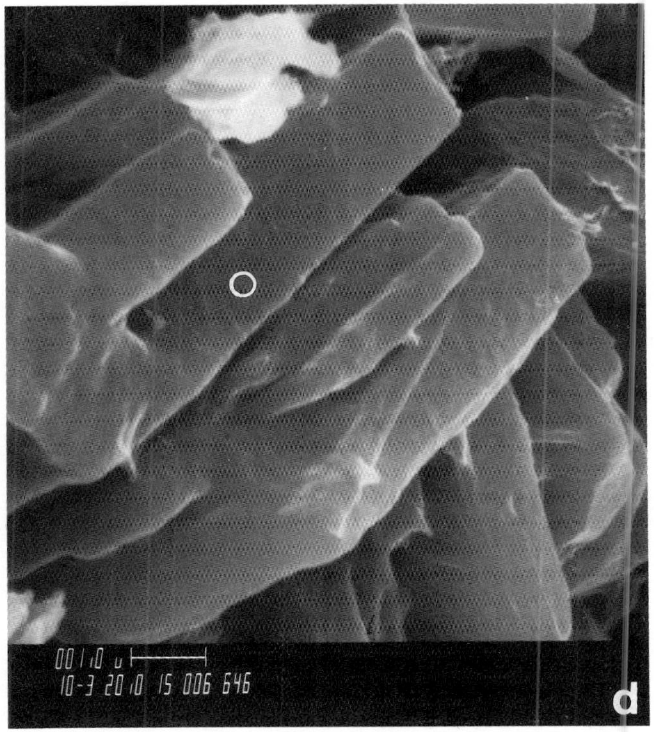

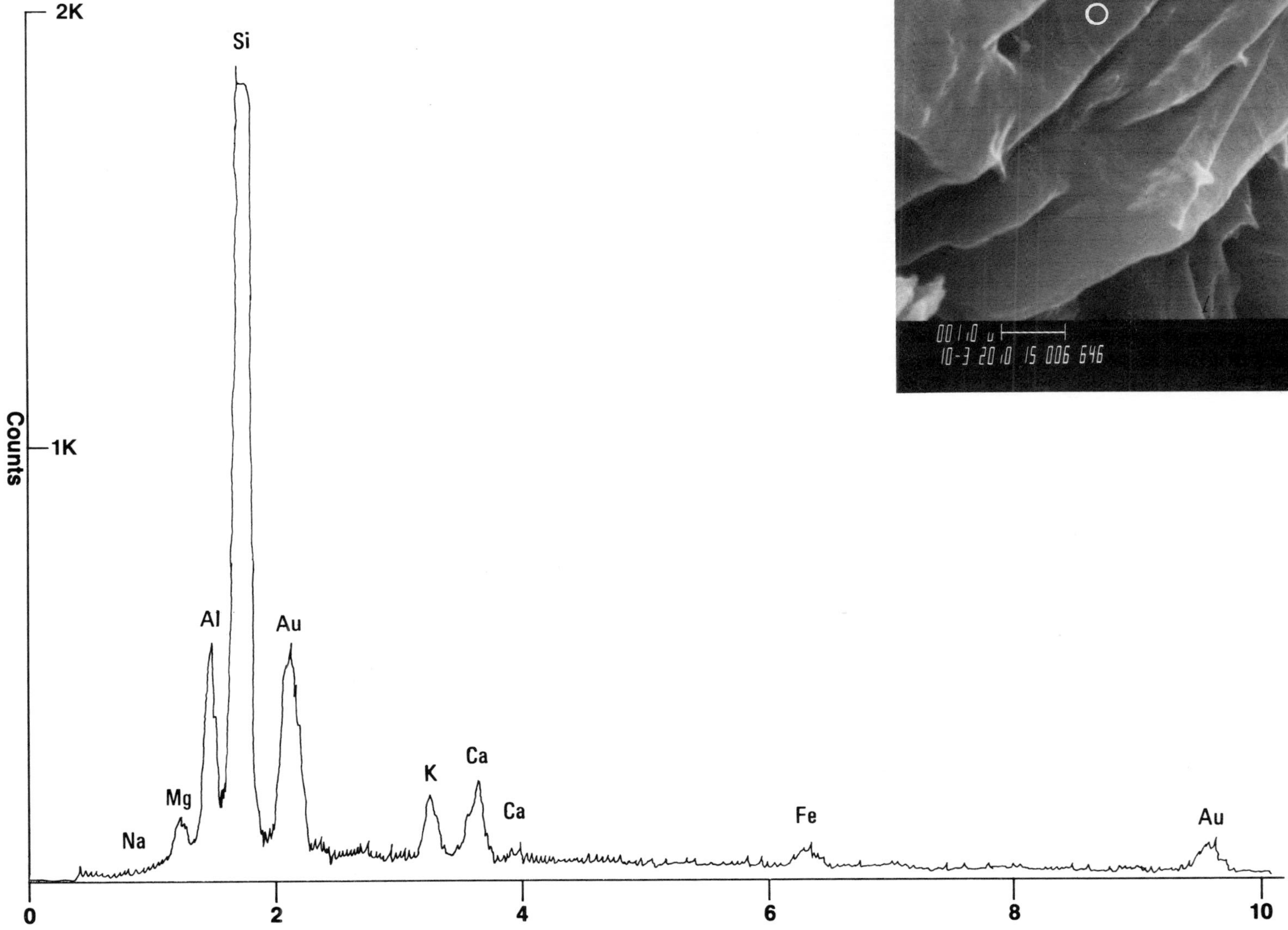

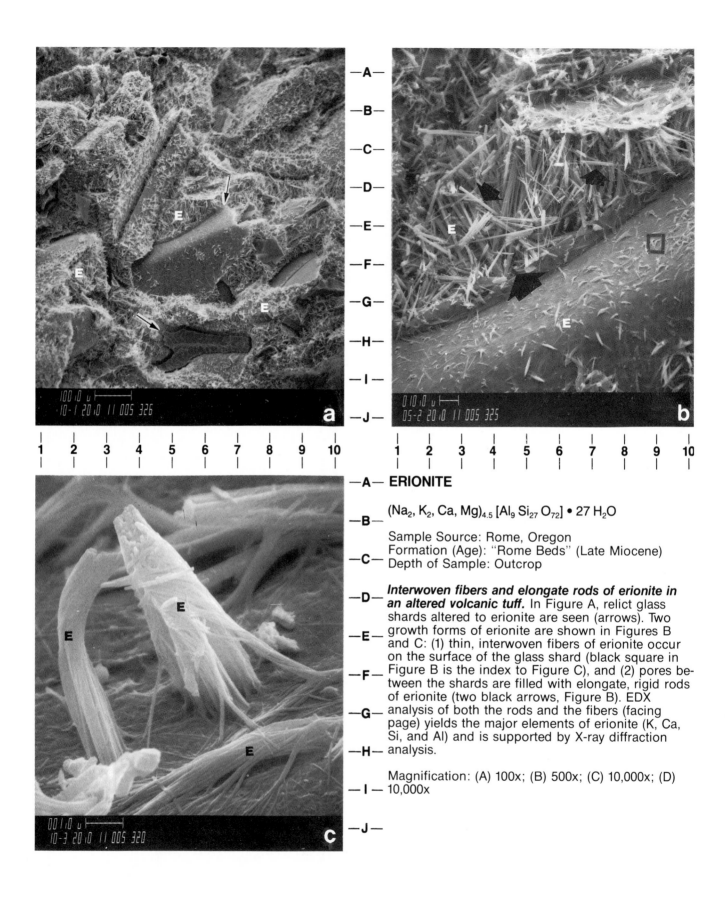

ERIONITE

$(Na_2, K_2, Ca, Mg)_{4.5} [Al_9 Si_{27} O_{72}] \bullet 27 H_2O$

Sample Source: Rome, Oregon
Formation (Age): "Rome Beds" (Late Miocene)
Depth of Sample: Outcrop

Interwoven fibers and elongate rods of erionite in an altered volcanic tuff. In Figure A, relict glass shards altered to erionite are seen (arrows). Two growth forms of erionite are shown in Figures B and C: (1) thin, interwoven fibers of erionite occur on the surface of the glass shard (black square in Figure B is the index to Figure C), and (2) pores between the shards are filled with elongate, rigid rods of erionite (two black arrows, Figure B). EDX analysis of both the rods and the fibers (facing page) yields the major elements of erionite (K, Ca, Si, and Al) and is supported by X-ray diffraction analysis.

Magnification: (A) 100x; (B) 500x; (C) 10,000x; (D) 10,000x

Energy Dispersive X—Ray Spectrum (EDX)

Erionite $(Na_2, K_2, Ca, Mg)_{4.5} [Al_9 Si_{27} O_{72}] \cdot 27 H_2O$

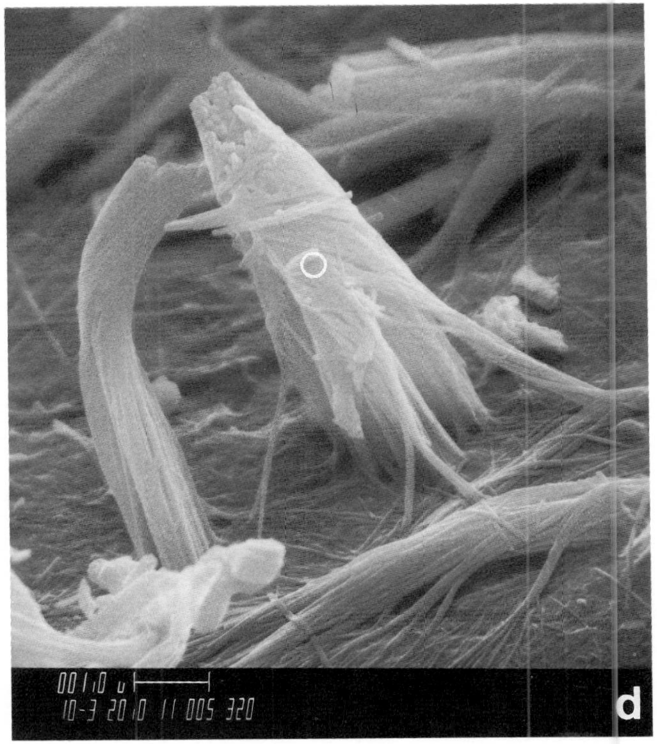

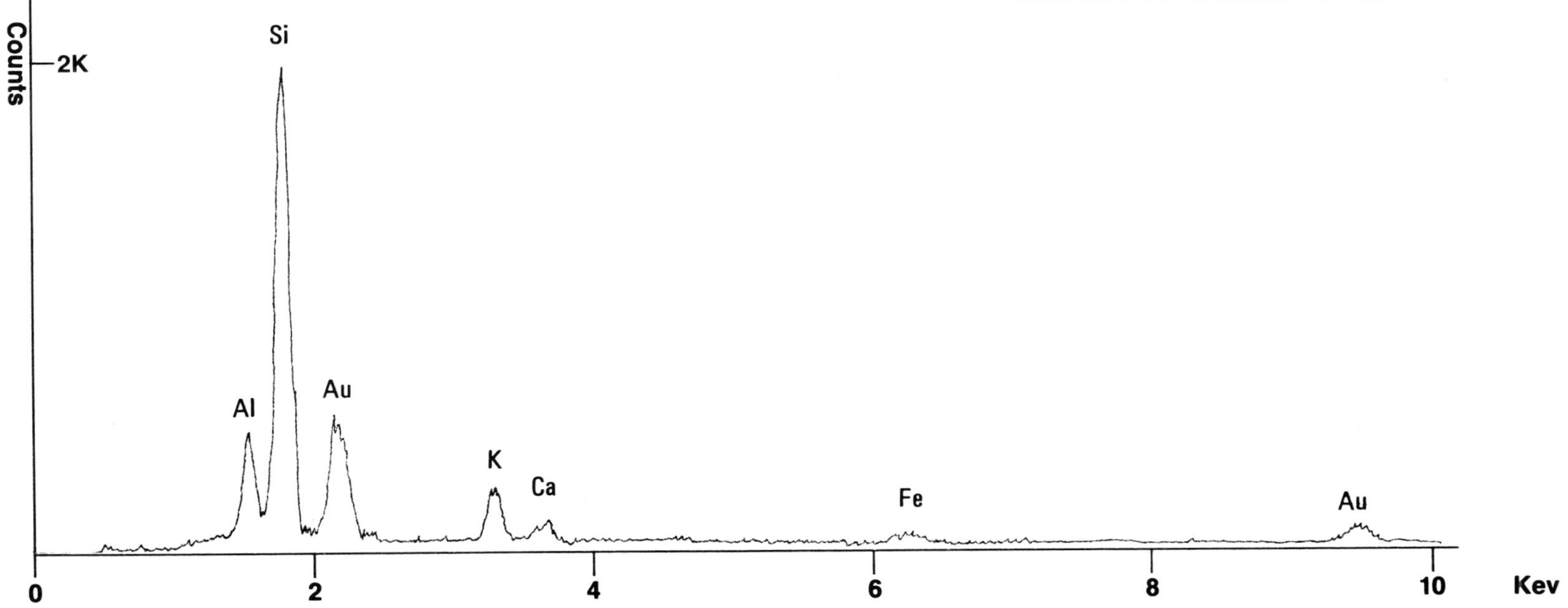

Silicates—Zeolite (Erionite)

119

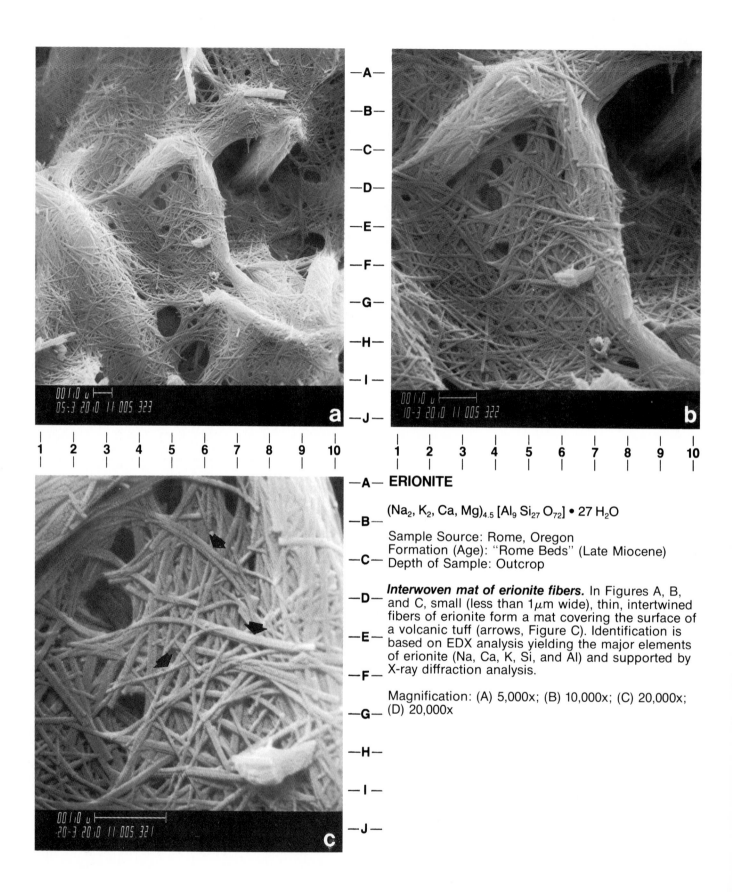

—A— **ERIONITE**

—B— $(Na_2, K_2, Ca, Mg)_{4.5} [Al_9 Si_{27} O_{72}] \bullet 27 H_2O$

Sample Source: Rome, Oregon
Formation (Age): "Rome Beds" (Late Miocene)
—C— Depth of Sample: Outcrop

Interwoven mat of erionite fibers. In Figures A, B, and C, small (less than 1μm wide), thin, intertwined fibers of erionite form a mat covering the surface of a volcanic tuff (arrows, Figure C). Identification is based on EDX analysis yielding the major elements of erionite (Na, Ca, K, Si, and Al) and supported by X-ray diffraction analysis.

Magnification: (A) 5,000x; (B) 10,000x; (C) 20,000x; (D) 20,000x

Energy Dispersive X—Ray Spectrum (EDX)

Erionite (Na_2, K_2, Ca, Mg)$_{4.5}$ [Al_9 Si_{27} O_{72}] • 27 H_2O

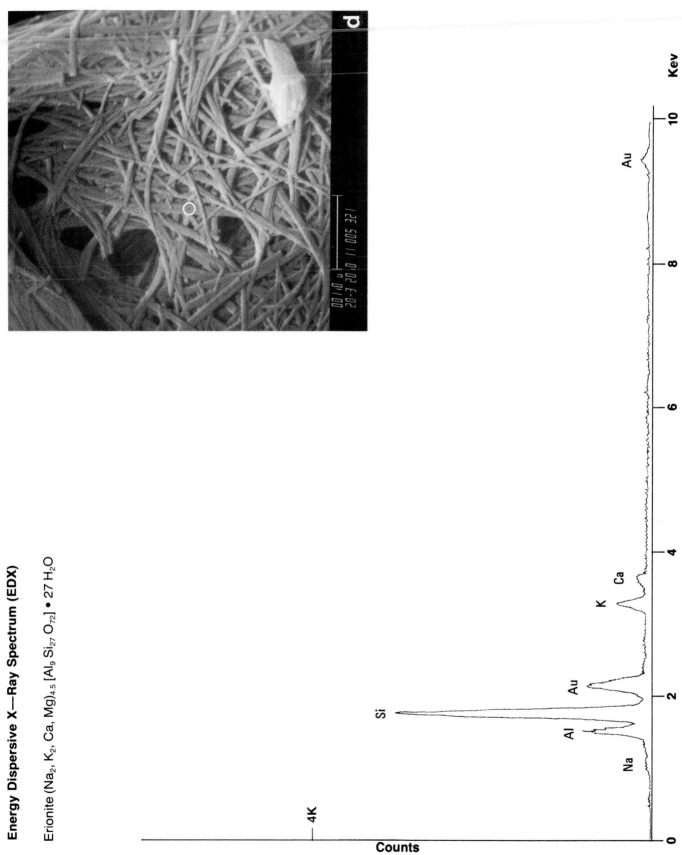

ERIONITE

$(Na_2,K_2,Ca,Mg)_{4.5} [Al_9Si_{27}O_{72}] \cdot 27 H_2O$

Sample Source: Bowie, Arizona
Formation (Age): (Pliocene-Pleistocene)
Depth of Formation: Outcrop

Barrel-shaped bundles of erionite surrounded by chabazite in a lacustrine tuff. In Figures A, B, and C, small (5 to 10μm) isolated bundles of erionite (E) are seen within a primarily chabazite tuff. Identification of these bundles as erionite is based on morphology and EDX analysis yielding the major elements of erionite (Na, K, Ca, Si, and Al). X-ray diffraction (XRD) analysis indicating the presence of erionite supports the identification.

Magnification: (A) 500x; (B) 1,000x; (C) 2,000x; (D) 2,000x

Silicates—Zeolite (Erionite)

Energy Dispersive X—Ray Spectrum (EDX)

Erionite $(Na_2, K_2, Ca, Mg)_{4.5}$ $[Al_9 Si_{27} O_{72}] \cdot 27 H_2O$

Silicates—Zeolite (Erionite)

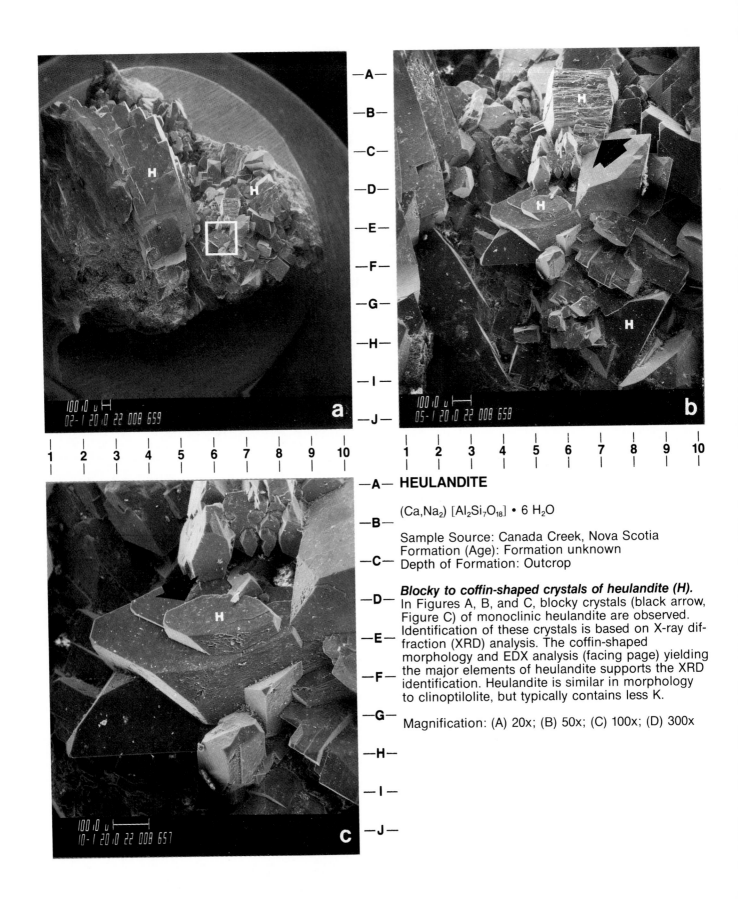

—A— **HEULANDITE**

—B— (Ca,Na_2) $[Al_2Si_7O_{18}]$ • 6 H_2O

Sample Source: Canada Creek, Nova Scotia
Formation (Age): Formation unknown
—C— Depth of Formation: Outcrop

—D— ***Blocky to coffin-shaped crystals of heulandite (H).***
In Figures A, B, and C, blocky crystals (black arrow,
Figure C) of monoclinic heulandite are observed.
Identification of these crystals is based on X-ray dif-
—E— fraction (XRD) analysis. The coffin-shaped
morphology and EDX analysis (facing page) yielding
—F— the major elements of heulandite supports the XRD
identification. Heulandite is similar in morphology
to clinoptilolite, but typically contains less K.

—G— Magnification: (A) 20x; (B) 50x; (C) 100x; (D) 300x

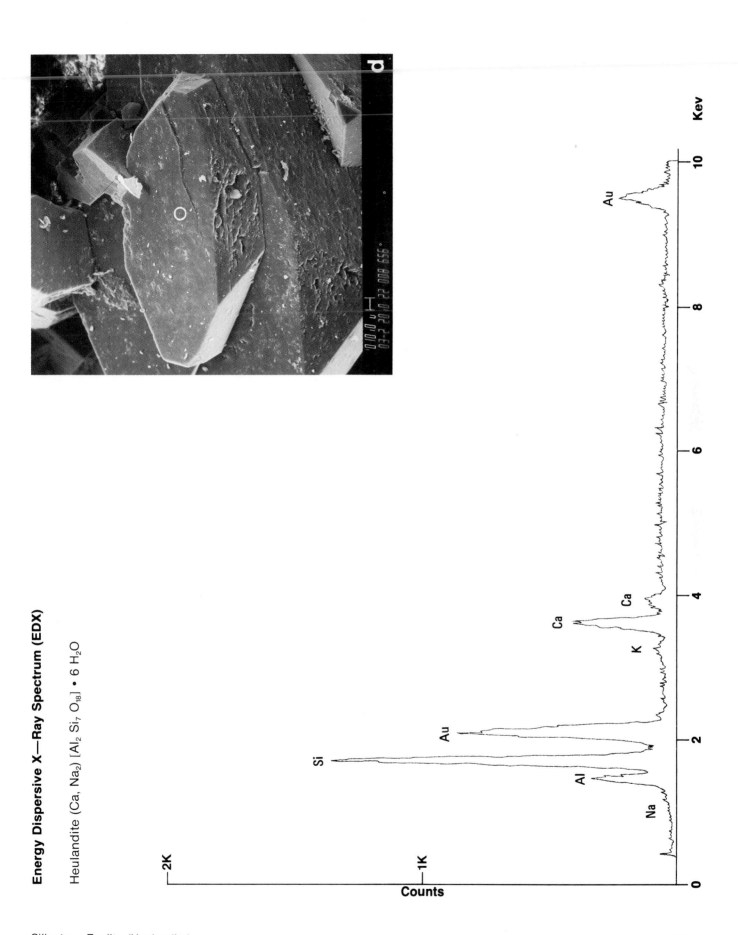

Energy Dispersive X—Ray Spectrum (EDX)

Heulandite (Ca, Na$_2$) [Al$_2$ Si$_7$ O$_{18}$] • 6 H$_2$O

Silicates—Zeolite (Heulandite)

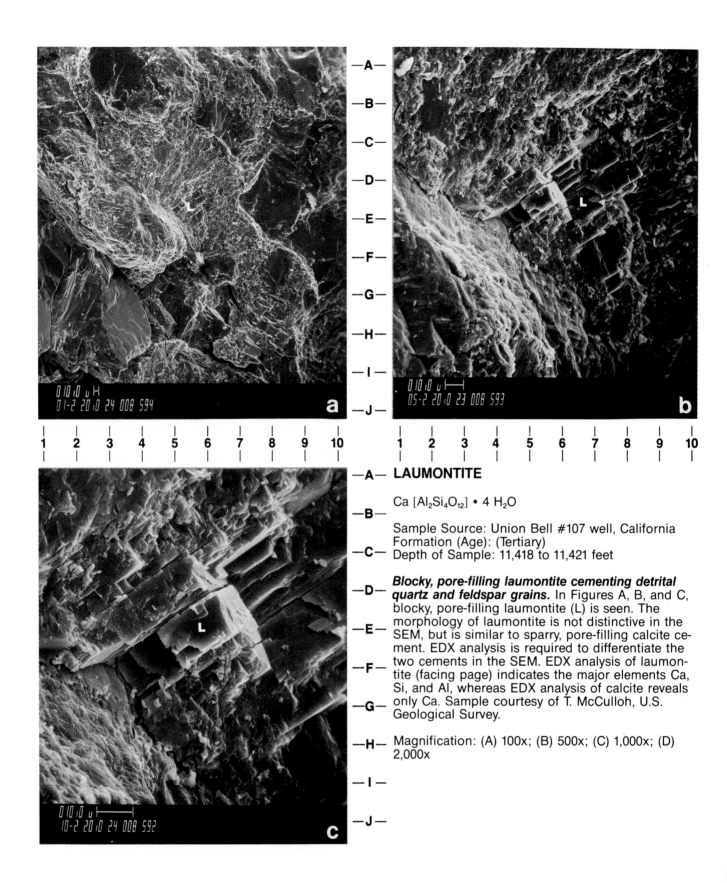

LAUMONTITE

Ca [Al₂Si₄O₁₂] • 4 H₂O

Sample Source: Union Bell #107 well, California
Formation (Age): (Tertiary)
Depth of Sample: 11,418 to 11,421 feet

Blocky, pore-filling laumontite cementing detrital quartz and feldspar grains. In Figures A, B, and C, blocky, pore-filling laumontite (L) is seen. The morphology of laumontite is not distinctive in the SEM, but is similar to sparry, pore-filling calcite cement. EDX analysis is required to differentiate the two cements in the SEM. EDX analysis of laumontite (facing page) indicates the major elements Ca, Si, and Al, whereas EDX analysis of calcite reveals only Ca. Sample courtesy of T. McCulloh, U.S. Geological Survey.

Magnification: (A) 100x; (B) 500x; (C) 1,000x; (D) 2,000x

Silicates—Zeolite (Laumontite)

Energy Dispersive X—Ray Spectrum (EDX)

Laumontite Ca [Al$_2$ Si$_4$ O$_{12}$] • 4 H$_2$O

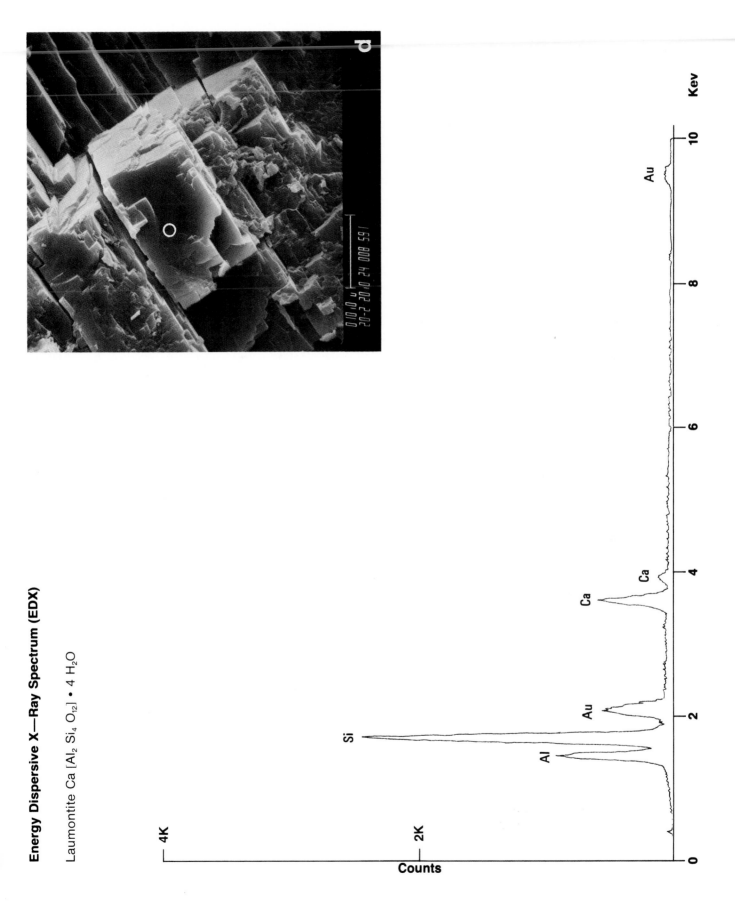

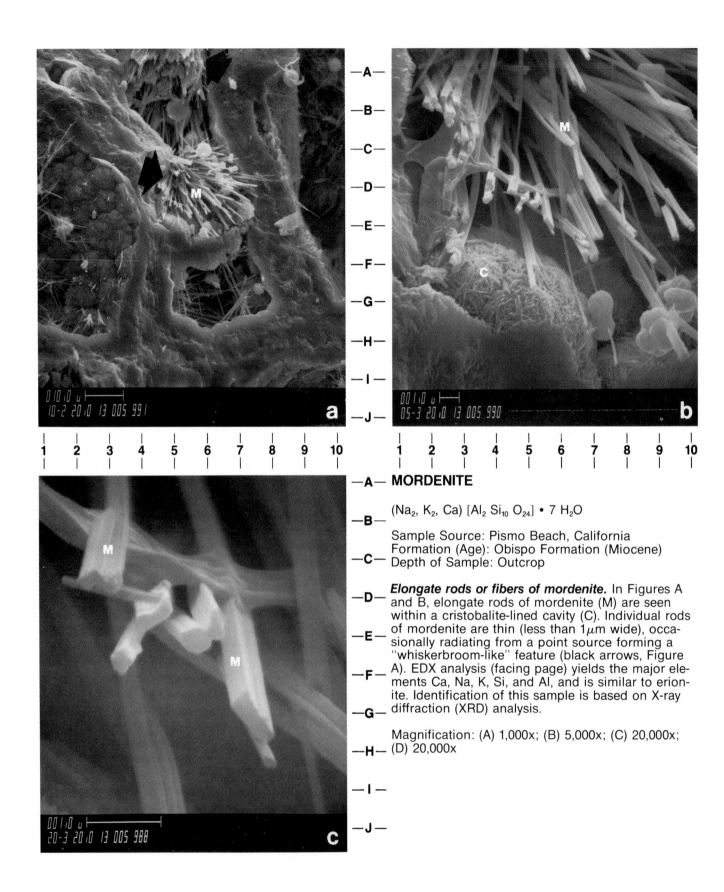

MORDENITE

(Na$_2$, K$_2$, Ca) [Al$_2$ Si$_{10}$ O$_{24}$] • 7 H$_2$O

Sample Source: Pismo Beach, California
Formation (Age): Obispo Formation (Miocene)
Depth of Sample: Outcrop

Elongate rods or fibers of mordenite. In Figures A and B, elongate rods of mordenite (M) are seen within a cristobalite-lined cavity (C). Individual rods of mordenite are thin (less than 1 μm wide), occasionally radiating from a point source forming a "whiskerbroom-like" feature (black arrows, Figure A). EDX analysis (facing page) yields the major elements Ca, Na, K, Si, and Al, and is similar to erionite. Identification of this sample is based on X-ray diffraction (XRD) analysis.

Magnification: (A) 1,000x; (B) 5,000x; (C) 20,000x; (D) 20,000x

128   Silicates—Zeolite (Mordenite)

Energy Dispersive X—Ray Spectrum (EDX)

Mordenite (Na$_2$, K$_2$, Ca) [Al$_2$ Si$_{10}$ O$_{24}$] • 7 H$_2$O

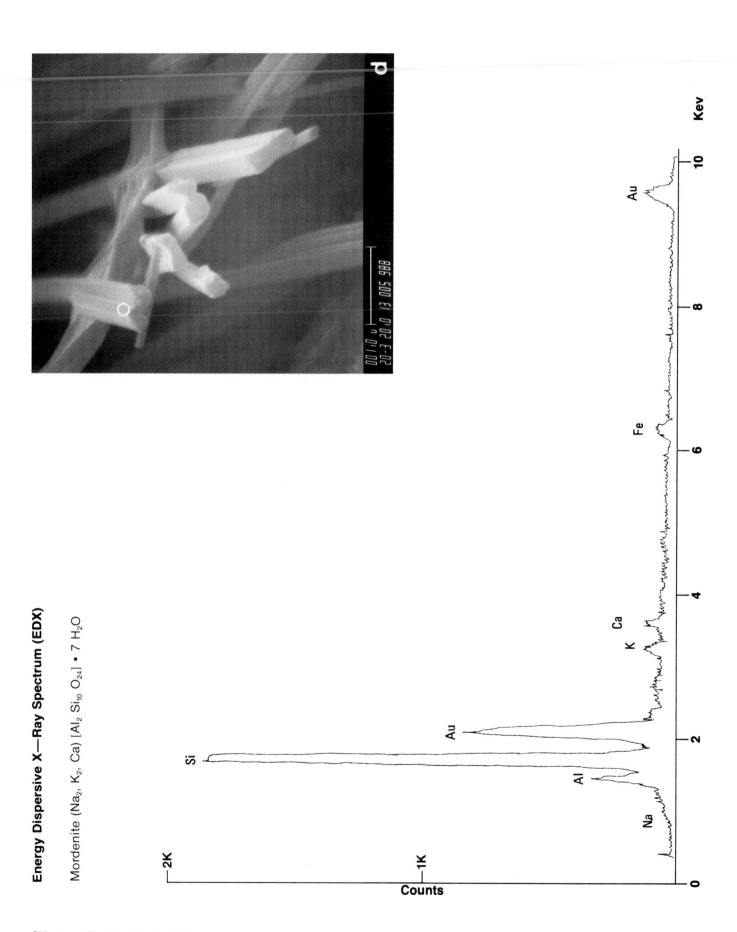

Silicates—Zeolite (Mordenite)

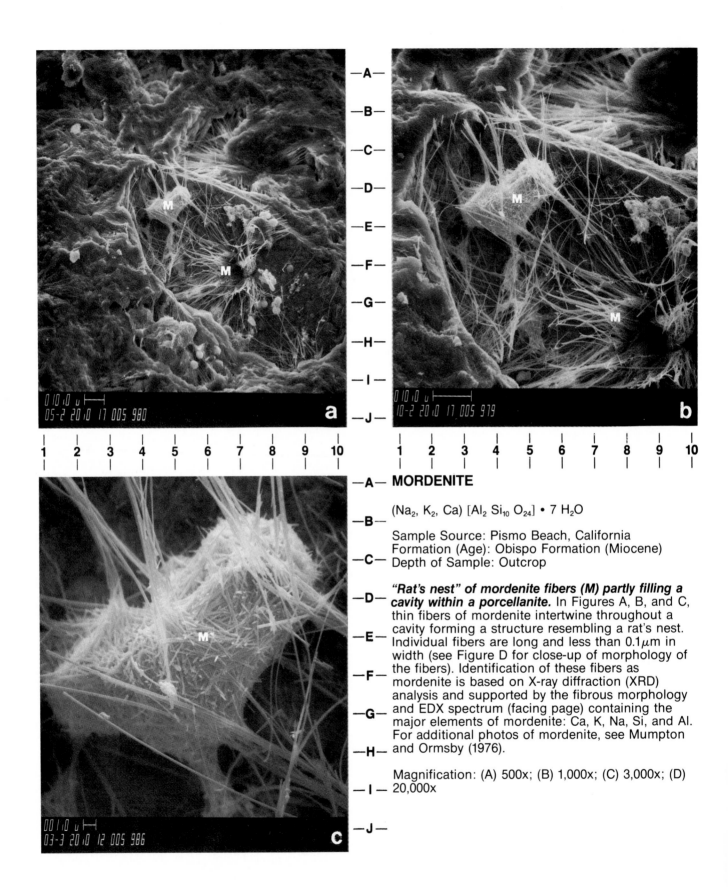

—A— **MORDENITE**

—B— $(Na_2, K_2, Ca) [Al_2 Si_{10} O_{24}] \cdot 7 H_2O$

—C— Sample Source: Pismo Beach, California
Formation (Age): Obispo Formation (Miocene)
Depth of Sample: Outcrop

—D— ***"Rat's nest" of mordenite fibers (M) partly filling a cavity within a porcellanite.*** In Figures A, B, and C, thin fibers of mordenite intertwine throughout a cavity forming a structure resembling a rat's nest.
—E— Individual fibers are long and less than $0.1\mu m$ in width (see Figure D for close-up of morphology of the fibers). Identification of these fibers as
—F— mordenite is based on X-ray diffraction (XRD) analysis and supported by the fibrous morphology and EDX spectrum (facing page) containing the
—G— major elements of mordenite: Ca, K, Na, Si, and Al. For additional photos of mordenite, see Mumpton and Ormsby (1976).
—H—

Magnification: (A) 500x; (B) 1,000x; (C) 3,000x; (D)
—I— 20,000x

—J—

Energy Dispersive X—Ray Spectrum (EDX)

Mordenite (Na$_2$, K$_2$, Ca) [Al$_2$ Si$_{10}$ O$_{24}$] • 7 H$_2$O

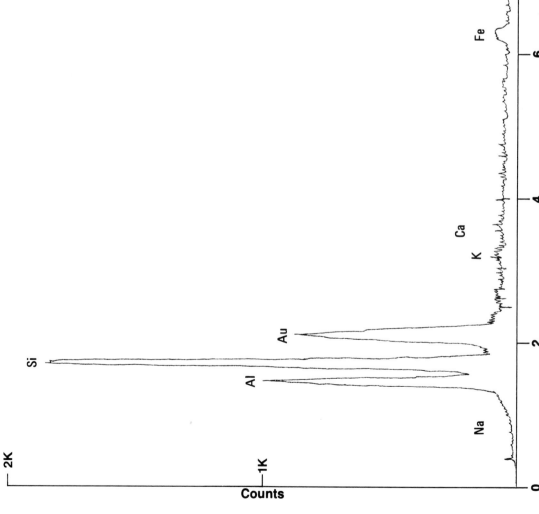

PHILLIPSITE

$(\tfrac{1}{2}Ca,Na,K)_3 \, [Al_3Si_5O_{16}] \cdot 6\,H_2O$

Sample Source: Shoshone, California
Formation (Age): Lake Tecopa Beds (Pleistocene)
Depth of Sample: Outcrop

Radiating clusters of phillipsite in a saline-lake tuff.
In Figures A, B, and C, stubby prisms of phillipsite (P), arranged into radiating clusters (Figure B, coordinates C3) or rosettes (Figure C) are seen. Individual crystals are small, ranging in size from 3 to $30\mu m$ in length and 0.3 to $3\mu m$ in width. Identification of these prisms as phillipsite is based on X-ray diffraction (XRD) analysis and supported by morphologic and EDX analysis (facing page) yielding the major elements of phillipsite: K, Na, Ca, Si, and Al.

Magnification: (A) 100x; (B) 600x; (C) 4,000x; (D) 4,000x

132

Energy Dispersive X—Ray Spectrum (EDX)

Phillipsite (½ Ca, Na, K)$_3$ [Al$_3$ Si$_5$ O$_{16}$] • 6 HO

Silicates—Zeolite (Phillipsite)

THOMSONITE

$NaCa_2 [(Al, Si)_5O_{10}]_2 \cdot 6 H_2O$

Sample Source: Goble, Oregon
Formation (Age): Goble Volcanics (Miocene)
Depth of Sample: Outcrop

Globular masses of thomsonite (T) lining vugs in a basalt. In Figures A and B, globular masses of radiating thomsonite crystals (arrows, Figure A), associated with blocky heulandite (Figure A, coordinates G5 and C5) are visible. A close-up of the surface of one of the balls (Figure C; black square in Figure B is the index) shows that the balls are composed of small (less than $0.1\mu m$ wide and 1 to $3\mu m$ long), tabular crystals oriented with the C-axis perpendicular to the growth surface (see Figure D for a close-up of the morphology). Identification of these crystals as thomsonite is based on X-ray diffraction (XRD) analysis and supported by EDX analysis (facing page) yielding the major elements of thomsonite: Ca, Na, Si, and Al. Sample courtesy of R. Beasley.

Magnification: (A) 30x; (B) 40x; (C) 800x; (D) 2,000x

Energy Dispersive X—Ray Spectrum (EDX)

Thomsonite Na Ca$_2$[(Al, Si)$_5$ O$_{10}$]$_2$ • 6 H$_2$O

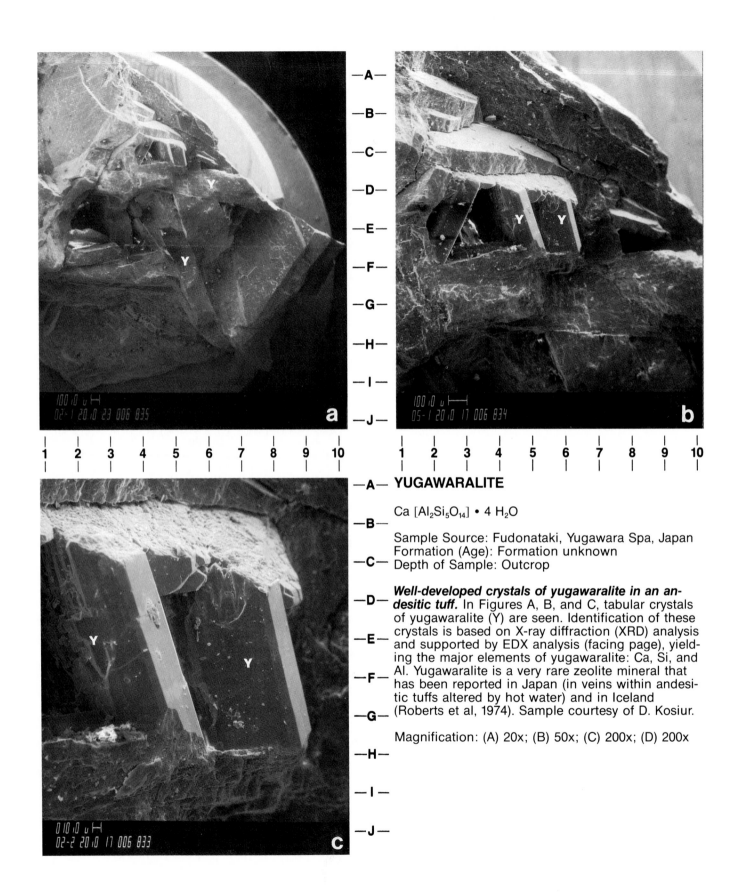

A B C D E F G H I J

1 2 3 4 5 6 7 8 9 10

1 2 3 4 5 6 7 8 9 10

YUGAWARALITE

Ca [Al$_2$Si$_5$O$_{14}$] • 4 H$_2$O

Sample Source: Fudonataki, Yugawara Spa, Japan
Formation (Age): Formation unknown
Depth of Sample: Outcrop

Well-developed crystals of yugawaralite in an andesitic tuff. In Figures A, B, and C, tabular crystals of yugawaralite (Y) are seen. Identification of these crystals is based on X-ray diffraction (XRD) analysis and supported by EDX analysis (facing page), yielding the major elements of yugawaralite: Ca, Si, and Al. Yugawaralite is a very rare zeolite mineral that has been reported in Japan (in veins within andesitic tuffs altered by hot water) and in Iceland (Roberts et al, 1974). Sample courtesy of D. Kosiur.

Magnification: (A) 20x; (B) 50x; (C) 200x; (D) 200x

Silicates—Zeolite (Yugawaralite)

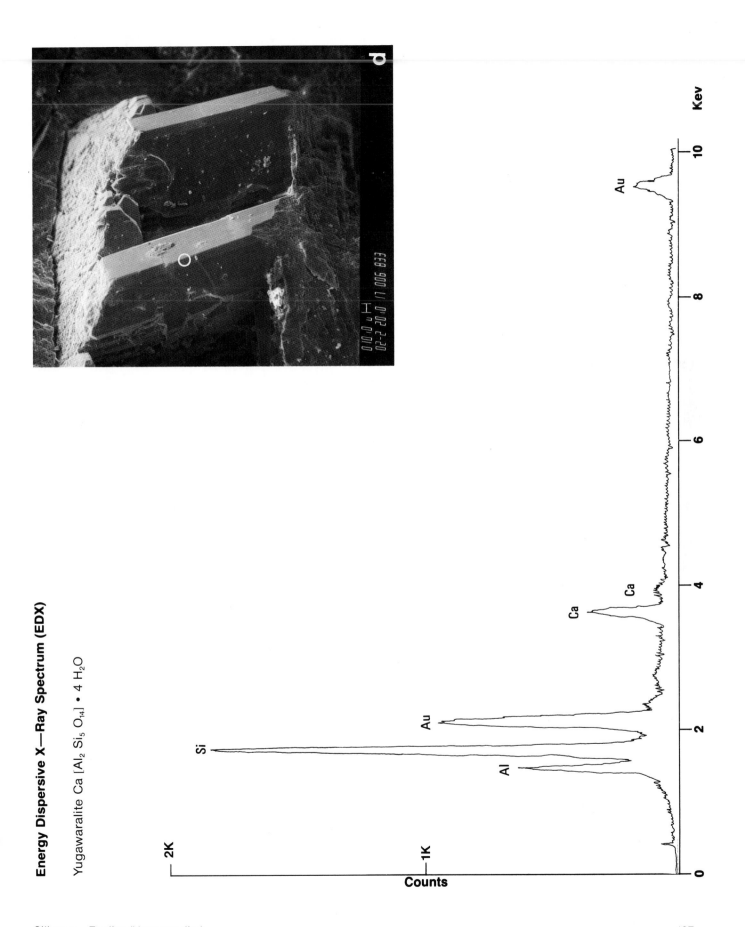

Energy Dispersive X—Ray Spectrum (EDX)

Yugawaralite Ca [Al$_2$ Si$_5$ O$_{14}$] • 4 H$_2$O

Counts

Kev

Micas

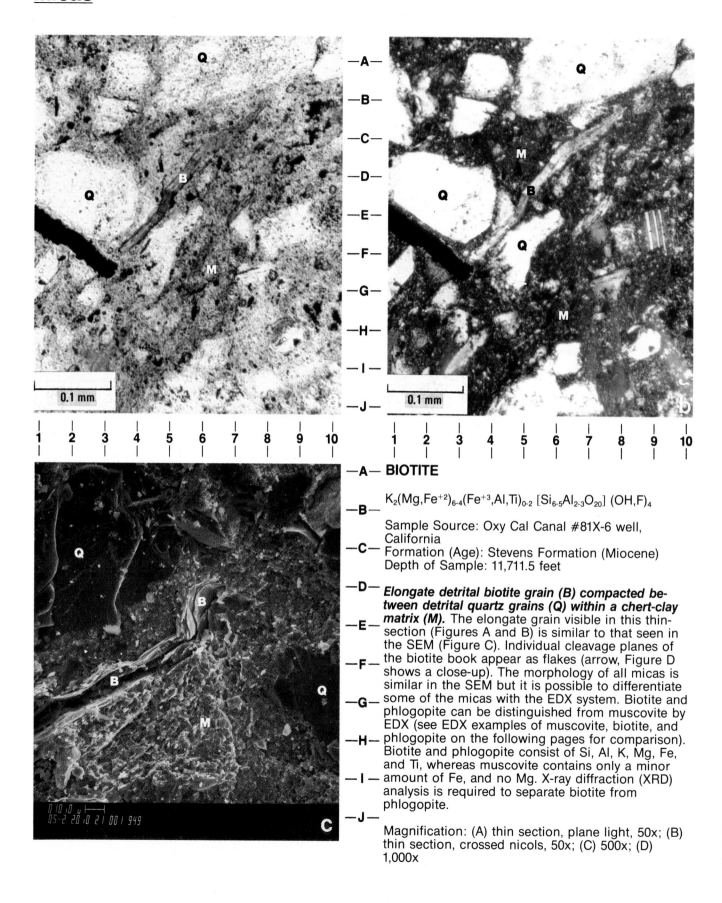

—A—	**BIOTITE**
—B—	$K_2(Mg,Fe^{+2})_{6-4}(Fe^{+3},Al,Ti)_{0-2}$ $[Si_{6-5}Al_{2-3}O_{20}]$ $(OH,F)_4$
—C—	Sample Source: Oxy Cal Canal #81X-6 well, California Formation (Age): Stevens Formation (Miocene) Depth of Sample: 11,711.5 feet
—D—	***Elongate detrital biotite grain (B) compacted between detrital quartz grains (Q) within a chert-clay matrix (M).*** The elongate grain visible in this thin-section (Figures A and B) is similar to that seen in
—E—	the SEM (Figure C). Individual cleavage planes of the biotite book appear as flakes (arrow, Figure D shows a close-up). The morphology of all micas is
—F—	similar in the SEM but it is possible to differentiate some of the micas with the EDX system. Biotite and phlogopite can be distinguished from muscovite by
—G—	EDX (see EDX examples of muscovite, biotite, and phlogopite on the following pages for comparison). Biotite and phlogopite consist of Si, Al, K, Mg, Fe,
—H—	and Ti, whereas muscovite contains only a minor amount of Fe, and no Mg. X-ray diffraction (XRD)
—I—	analysis is required to separate biotite from phlogopite.
—J—	Magnification: (A) thin section, plane light, 50x; (B) thin section, crossed nicols, 50x; (C) 500x; (D) 1,000x

Silicates—Mica (Biotite)

Energy Dispersive X—Ray Spectrum (EDX)

Biotite K_2 (Mg, Fe^{+2})$_{6-4}$ (Fe^{+3}, Al, Ti)$_{0-2}$ [Si_{6-5} Al_{2-3} O_{20}] (OH, F)$_4$

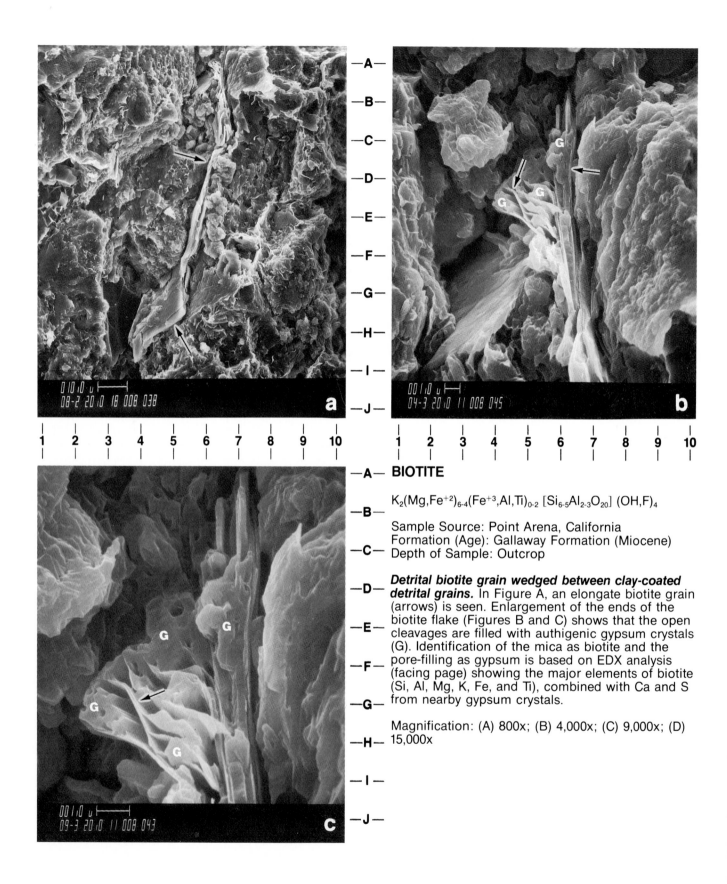

BIOTITE

$K_2(Mg,Fe^{+2})_{6-4}(Fe^{+3},Al,Ti)_{0-2}[Si_{6-5}Al_{2-3}O_{20}](OH,F)_4$

Sample Source: Point Arena, California
Formation (Age): Gallaway Formation (Miocene)
Depth of Sample: Outcrop

Detrital biotite grain wedged between clay-coated detrital grains. In Figure A, an elongate biotite grain (arrows) is seen. Enlargement of the ends of the biotite flake (Figures B and C) shows that the open cleavages are filled with authigenic gypsum crystals (G). Identification of the mica as biotite and the pore-filling as gypsum is based on EDX analysis (facing page) showing the major elements of biotite (Si, Al, Mg, K, Fe, and Ti), combined with Ca and S from nearby gypsum crystals.

Magnification: (A) 800x; (B) 4,000x; (C) 9,000x; (D) 15,000x

Energy Dispersive X—Ray Spectrum (EDX)

Biotite K_2 (Mg, Fe^{+2}) $_{6-4}$ (Fe^{+3}, Al, Ti) $_{0-2}$ [Si_{6-5} Al_{2-3} O_{20}] (OH, F) $_4$

Energy Dispersive X—Ray Spectrum (EDX)

Muscovite $K_2 Al_4 [Si_6 Al_2 O_{20}] (OH, F)_4$

Silicates—Mica (Muscovite)

Energy Dispersive X—Ray Spectrum (EDX)

Muscovite $K_2 Al_4 [Si_6 Al_2 O_{20}] (OH, F)_4$

Energy Dispersive X—Ray Spectrum (EDX)

Phlogopite $K_2 (Mg, Fe^{+2})_6 [Si_6 Al_2 O_{20}] (OH, F)_4$

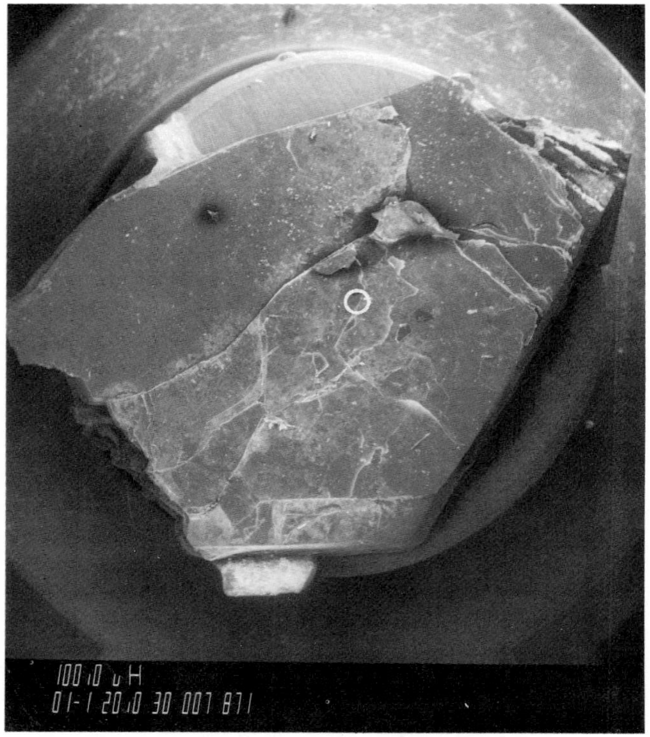

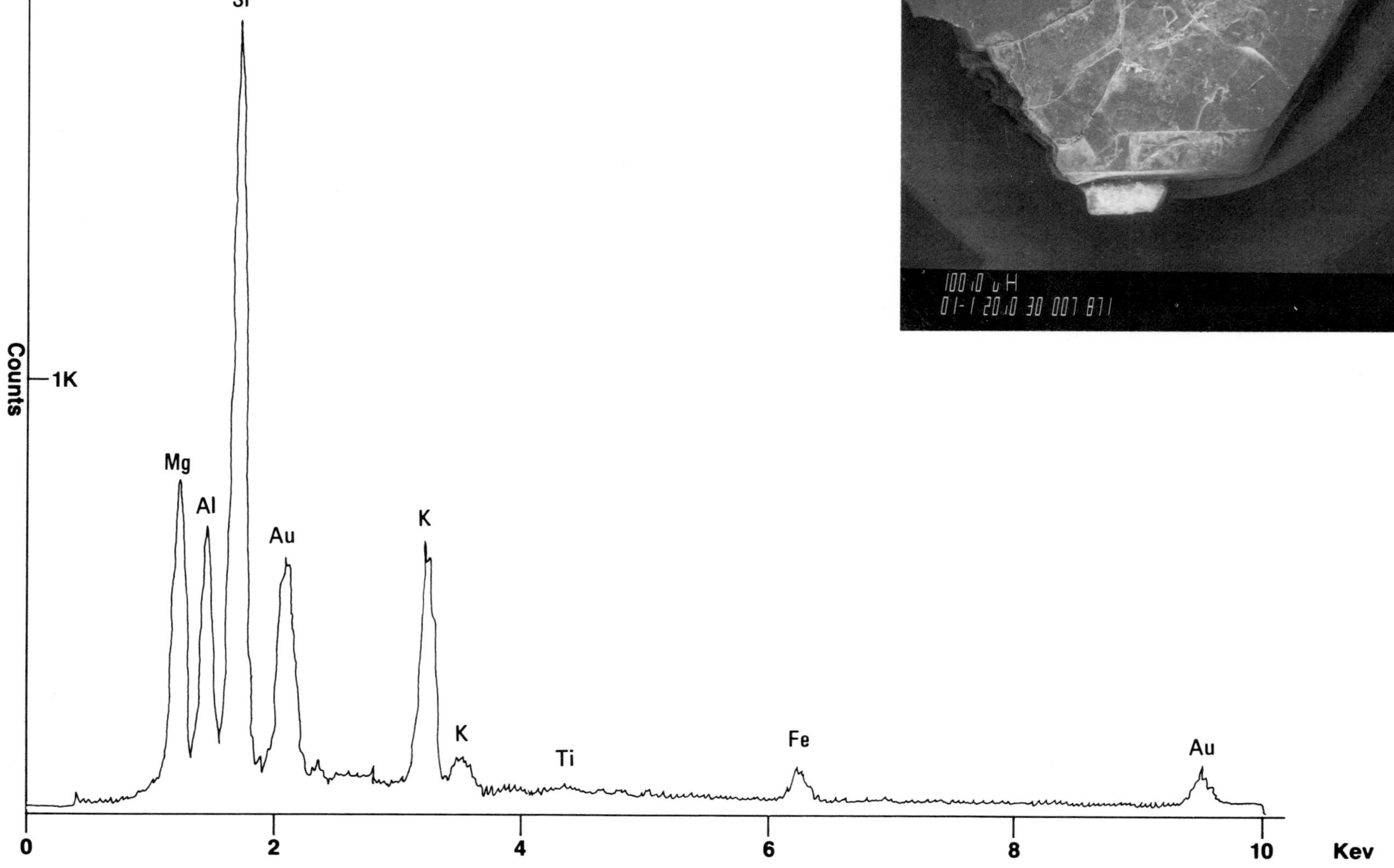

Energy Dispersive X—Ray Spectrum (EDX)

Hornblende $(Na, K)_{0-1} Ca_2 (Mg, Fe^{+2}, Fe^{+3}, Al)_5 [Si_{6-7} Al_{2-1} O_{22}] (OH, F)_2$

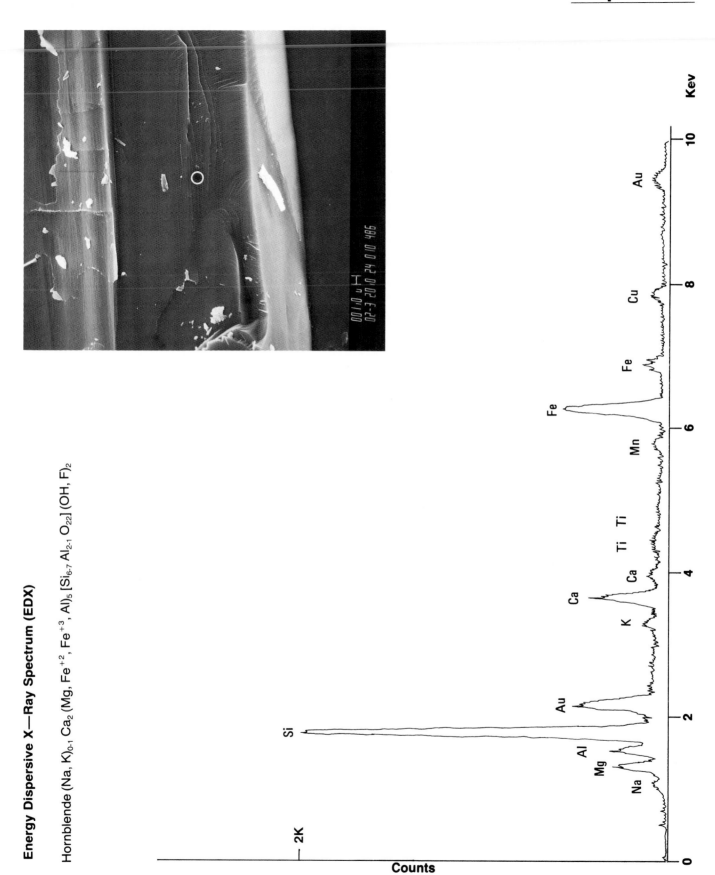

Silicates—Amphibole (Hornblende)

Energy Dispersive X—Ray Spectrum (EDX)

Actinolite Ca_2 $(Mg, Fe^{+2})_5$ $[Si_8 O_{22}]$ $(OH, F)_2$

Silicates—Amphibole (Actinolite)

Energy Dispersive X—Ray Spectrum (EDX)

Riebeckite $Na_2 Fe_3^{+2} Fe_2^{+3} [Si_8 O_{22}] (OH)_2$

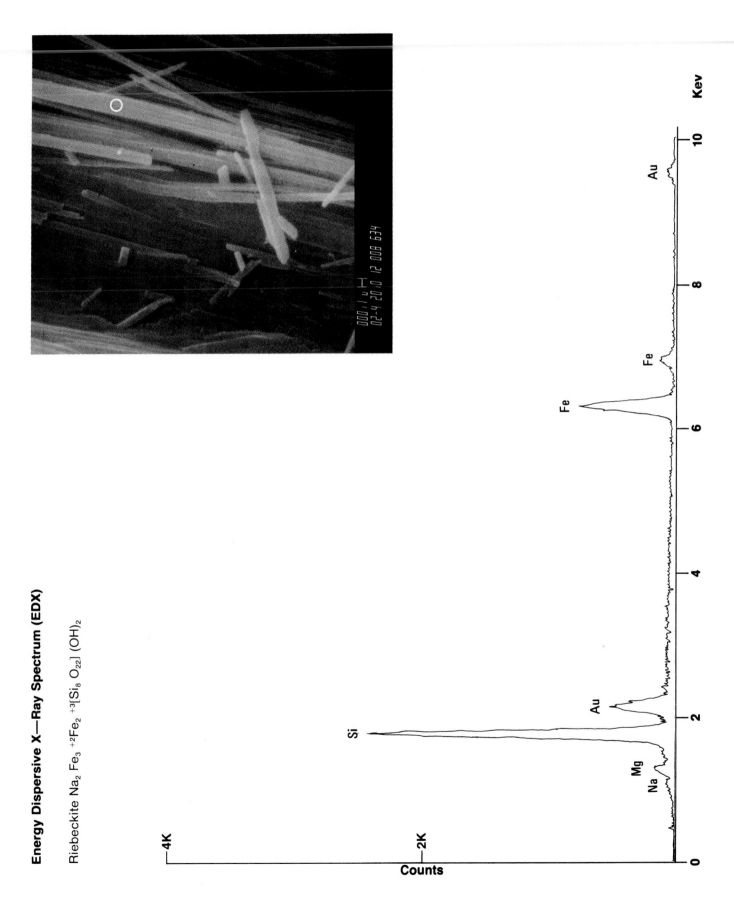

Pyroxenes

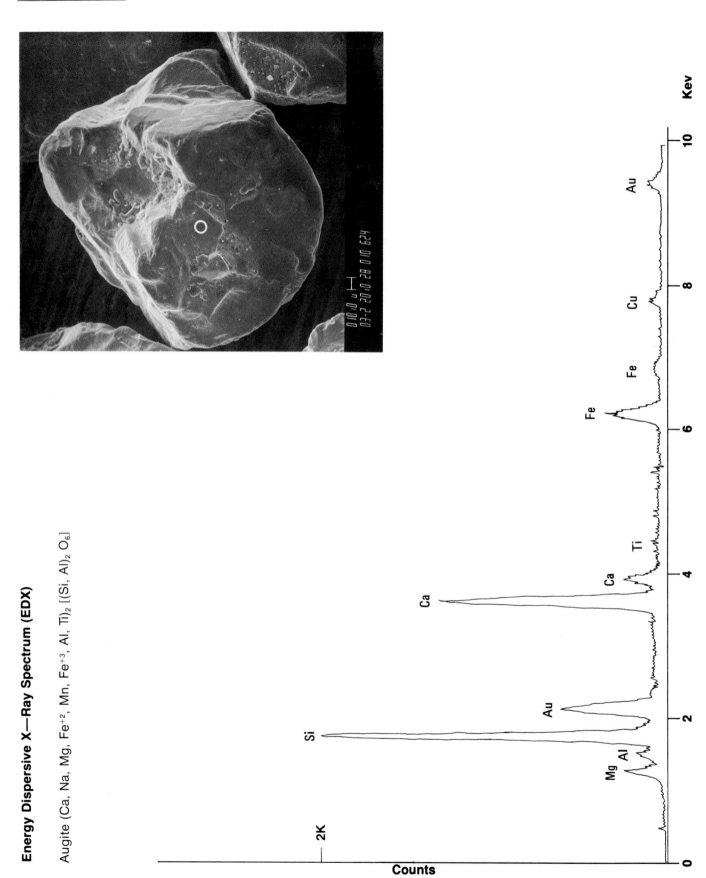

Energy Dispersive X—Ray Spectrum (EDX)

Augite (Ca, Na, Mg, Fe^{+2}, Mn, Fe^{+3}, Al, Ti)$_2$ [(Si, Al)$_2$ O$_6$]

Silicates—Pyroxene (Augite)

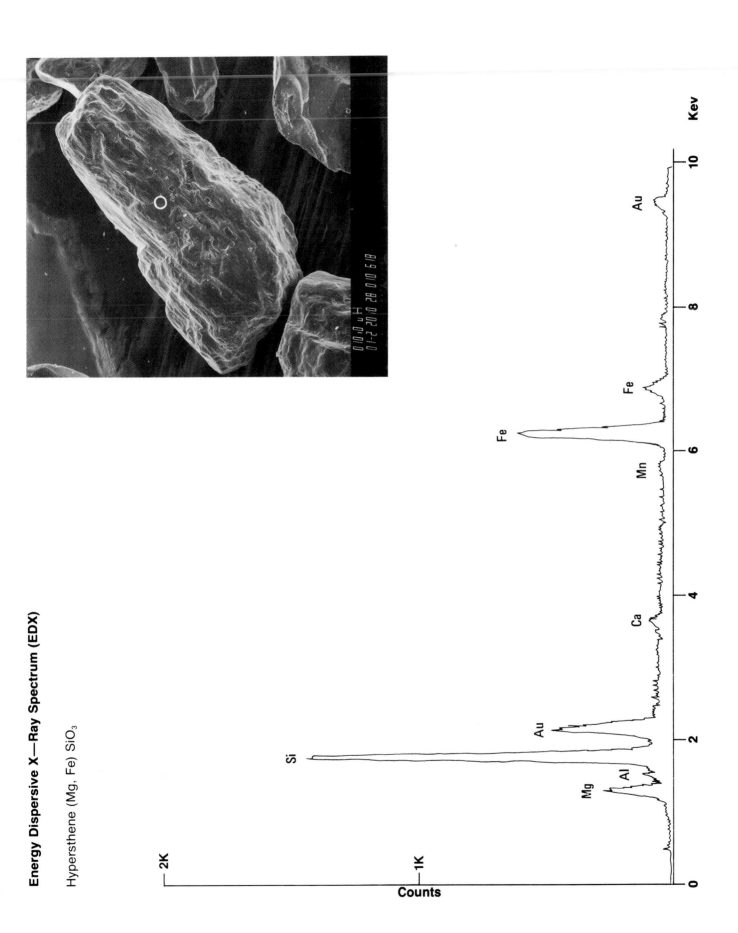

Energy Dispersive X—Ray Spectrum (EDX)

Hypersthene (Mg, Fe) SiO_3

Silicates—Pyroxene (Hypersthene)

Others

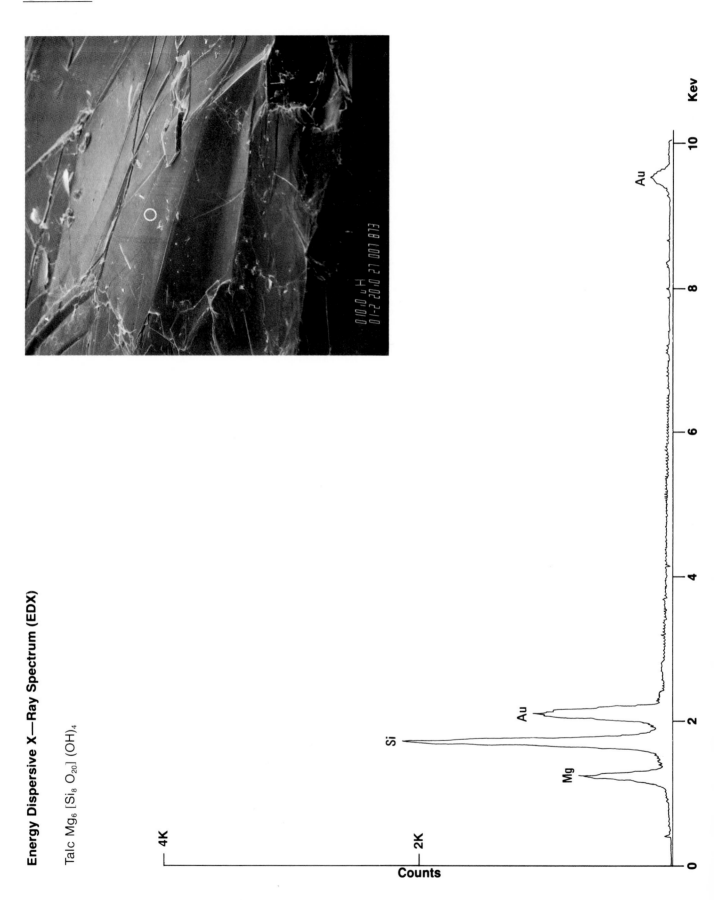

Energy Dispersive X—Ray Spectrum (EDX)

Talc $Mg_6 [Si_8 O_{20}] (OH)_4$

150

Silicates—Talc

Energy Dispersive X—Ray Spectrum (EDX)

Chrysotile $Mg_3 [Si_2 O_5] (OH)_4$

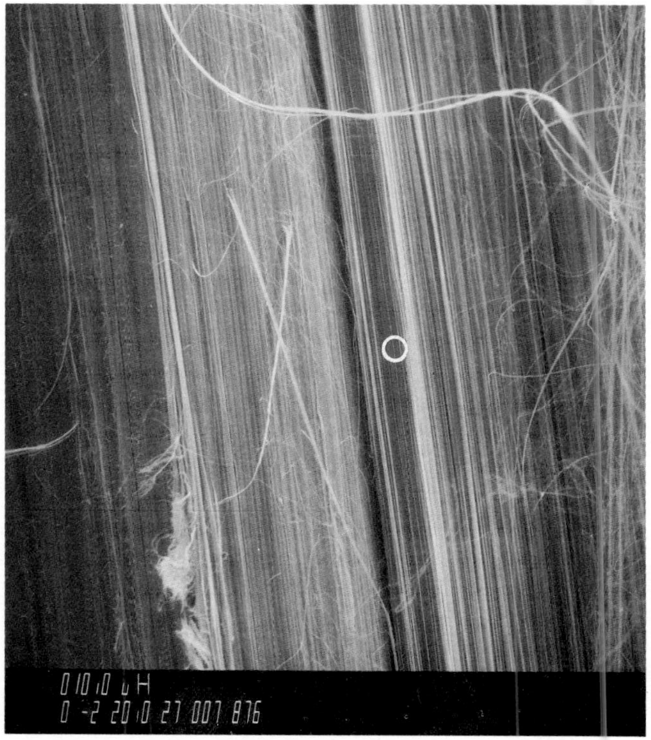

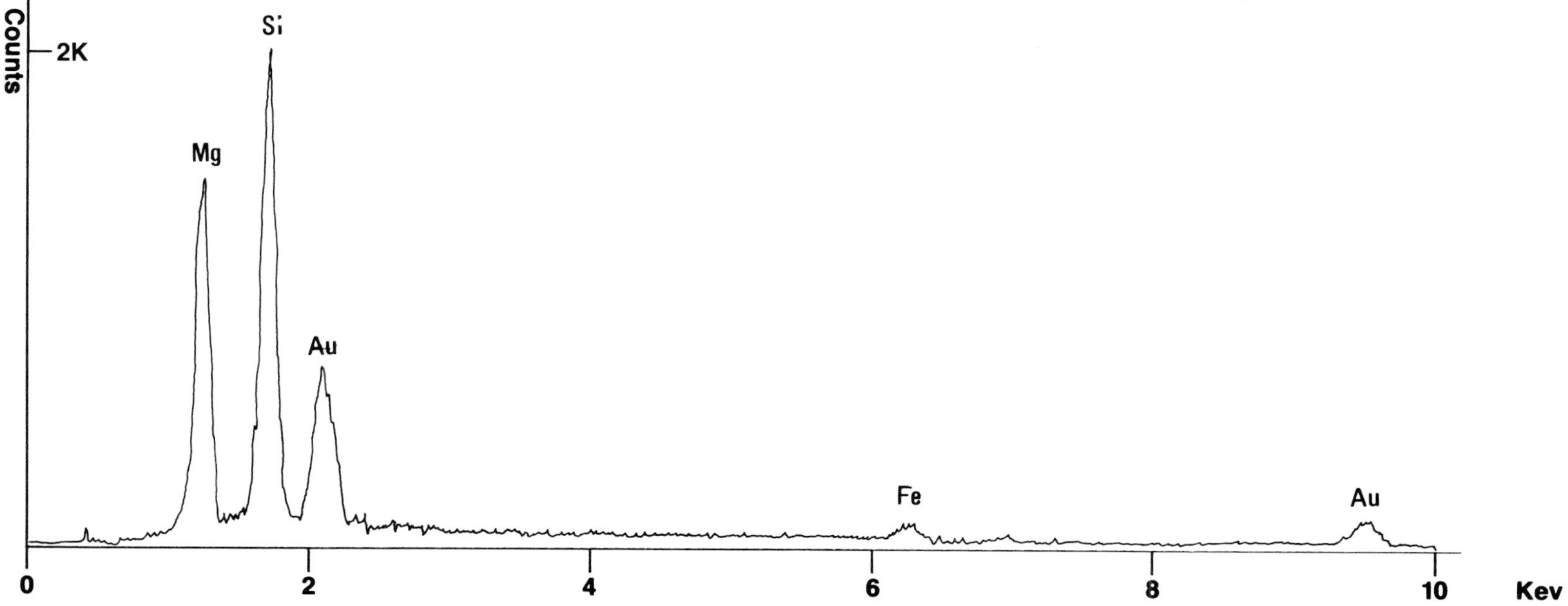

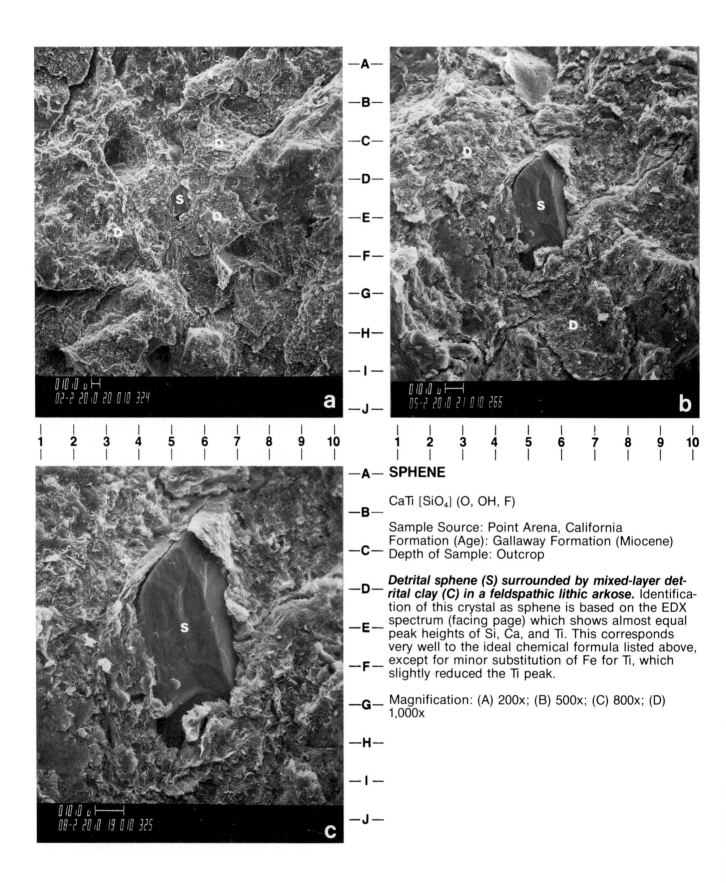

SPHENE

CaTi [SiO$_4$] (O, OH, F)

Sample Source: Point Arena, California
Formation (Age): Gallaway Formation (Miocene)
Depth of Sample: Outcrop

Detrital sphene (S) surrounded by mixed-layer detrital clay (C) in a feldspathic lithic arkose. Identification of this crystal as sphene is based on the EDX spectrum (facing page) which shows almost equal peak heights of Si, Ca, and Ti. This corresponds very well to the ideal chemical formula listed above, except for minor substitution of Fe for Ti, which slightly reduced the Ti peak.

Magnification: (A) 200x; (B) 500x; (C) 800x; (D) 1,000x

Energy Dispersive X—Ray Spectrum (EDX)

Sphene CaTi [SiO$_4$] (O, OH, F)

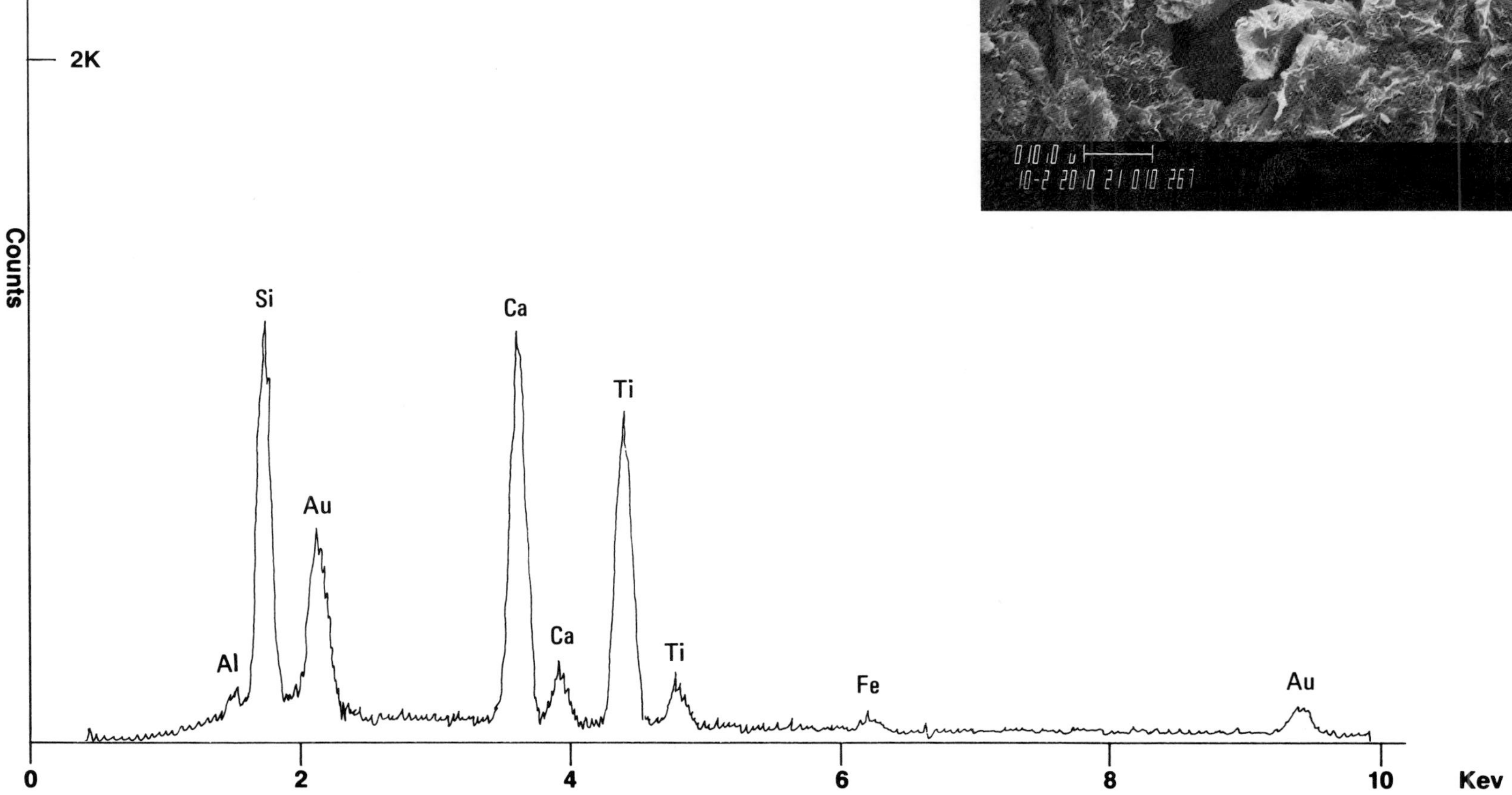

Silicates—Sphene

153

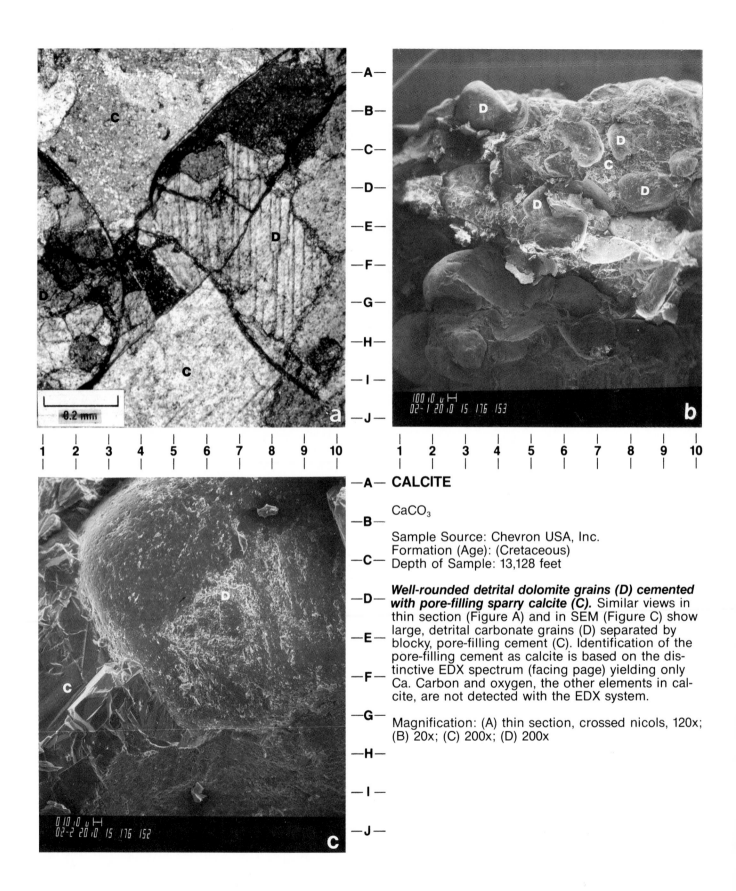

A B C D E F G H I J

| 1 | 2 | 3 | 4 | 5 | 6 | 7 | 8 | 9 | 10 |

CALCITE

CaCO$_3$

Sample Source: Chevron USA, Inc.
Formation (Age): (Cretaceous)
Depth of Sample: 13,128 feet

Well-rounded detrital dolomite grains (D) cemented with pore-filling sparry calcite (C). Similar views in thin section (Figure A) and in SEM (Figure C) show large, detrital carbonate grains (D) separated by blocky, pore-filling cement (C). Identification of the pore-filling cement as calcite is based on the distinctive EDX spectrum (facing page) yielding only Ca. Carbon and oxygen, the other elements in calcite, are not detected with the EDX system.

Magnification: (A) thin section, crossed nicols, 120x; (B) 20x; (C) 200x; (D) 200x

Energy Dispersive X—Ray Spectrum (EDX)

Calcite Ca CO₃

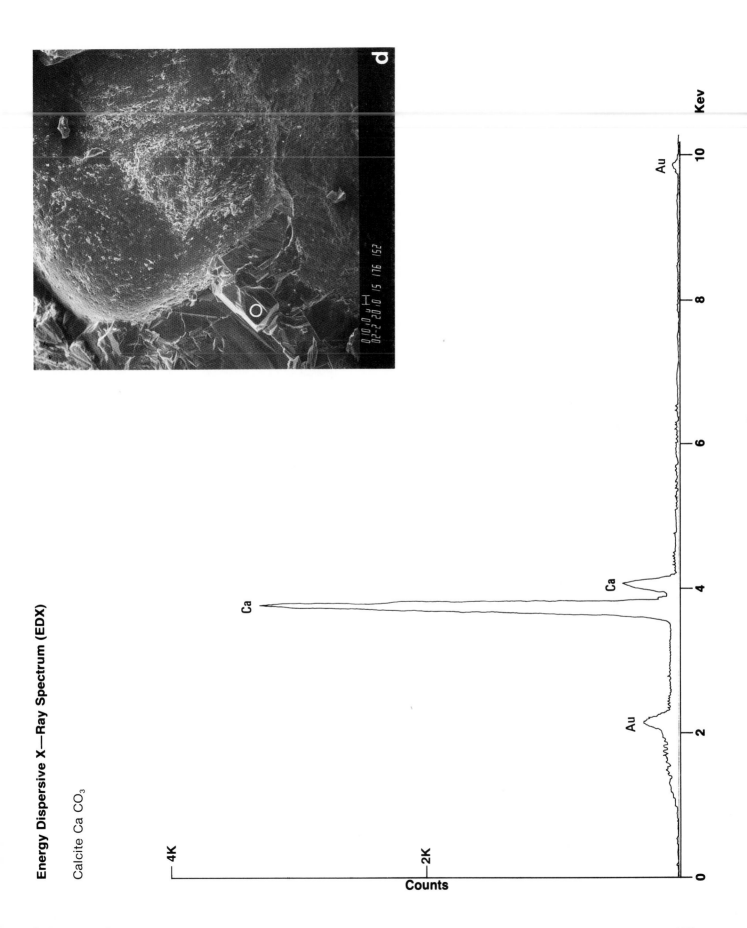

Counts

4K

2K

Ca

Ca

Au

Au

0   2   4   6   8   10   Kev

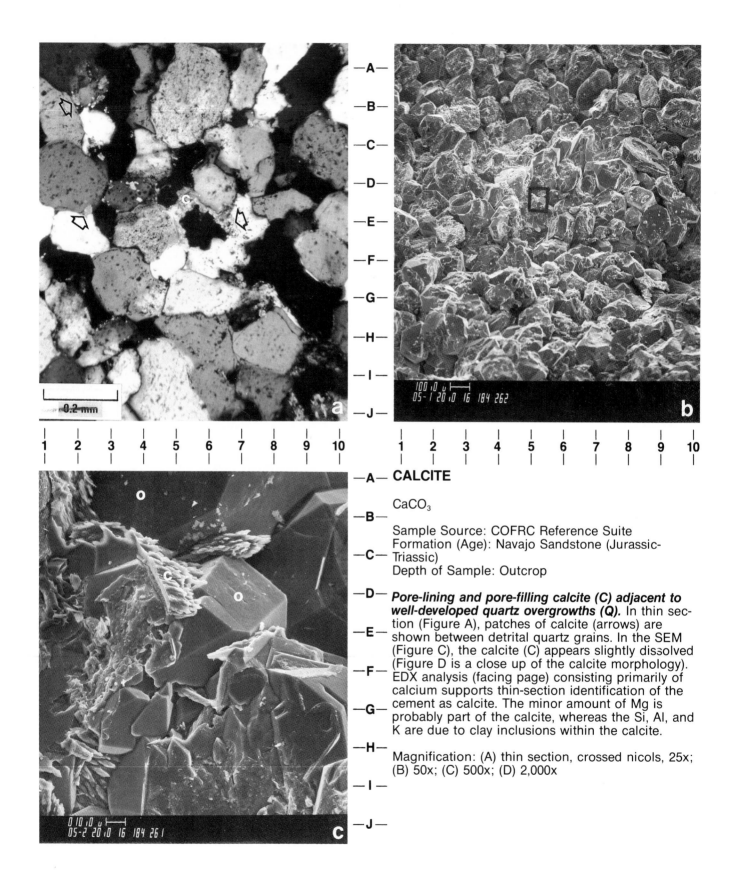

CALCITE

CaCO$_3$

Sample Source: COFRC Reference Suite
Formation (Age): Navajo Sandstone (Jurassic-Triassic)
Depth of Sample: Outcrop

Pore-lining and pore-filling calcite (C) adjacent to well-developed quartz overgrowths (Q). In thin section (Figure A), patches of calcite (arrows) are shown between detrital quartz grains. In the SEM (Figure C), the calcite (C) appears slightly dissolved (Figure D is a close up of the calcite morphology). EDX analysis (facing page) consisting primarily of calcium supports thin-section identification of the cement as calcite. The minor amount of Mg is probably part of the calcite, whereas the Si, Al, and K are due to clay inclusions within the calcite.

Magnification: (A) thin section, crossed nicols, 25x; (B) 50x; (C) 500x; (D) 2,000x

Energy Dispersive X—Ray Spectrum (EDX)

Calcite Ca CO$_3$

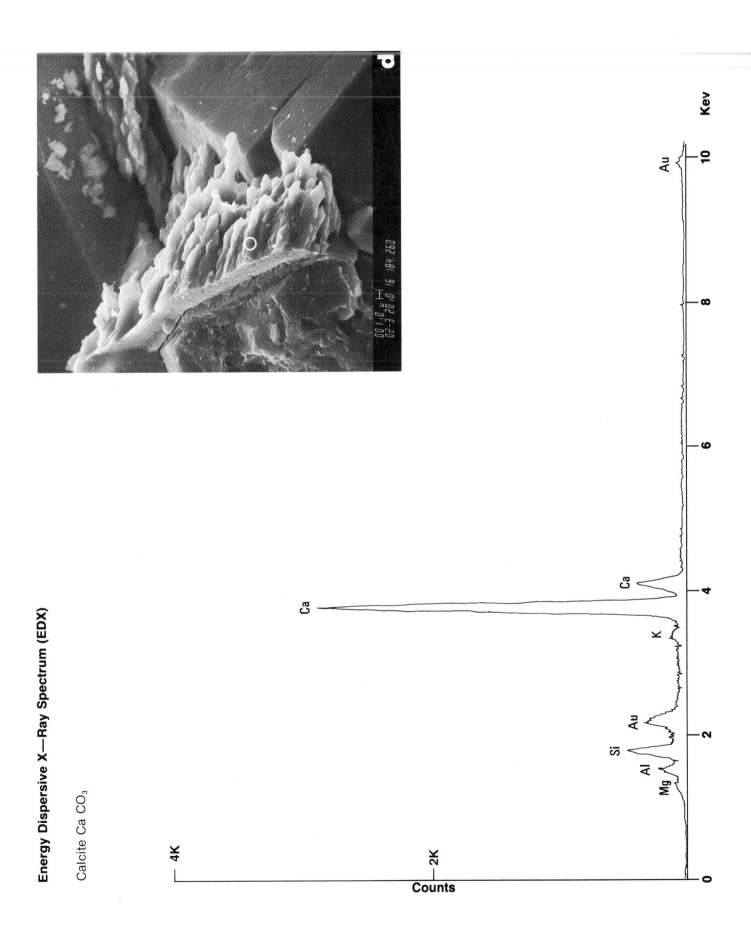

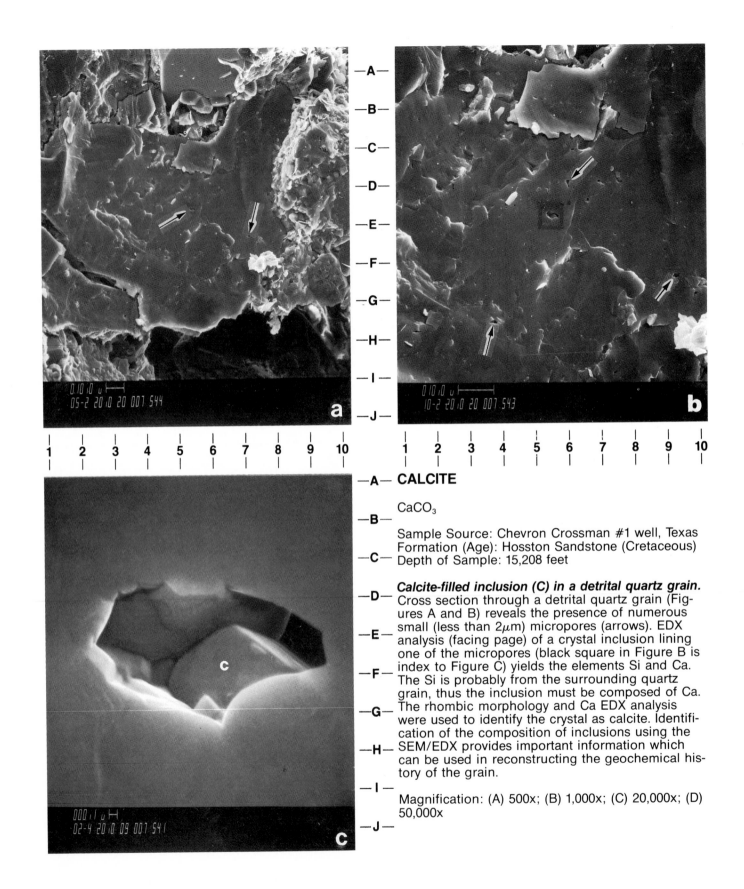

—A— **CALCITE**

—B— CaCO$_3$

—C— Sample Source: Chevron Crossman #1 well, Texas
Formation (Age): Hosston Sandstone (Cretaceous)
Depth of Sample: 15,208 feet

—D— ***Calcite-filled inclusion (C) in a detrital quartz grain.***
Cross section through a detrital quartz grain (Figures A and B) reveals the presence of numerous small (less than 2μm) micropores (arrows). EDX analysis (facing page) of a crystal inclusion lining one of the micropores (black square in Figure B is index to Figure C) yields the elements Si and Ca. The Si is probably from the surrounding quartz grain, thus the inclusion must be composed of Ca. The rhombic morphology and Ca EDX analysis were used to identify the crystal as calcite. Identification of the composition of inclusions using the SEM/EDX provides important information which can be used in reconstructing the geochemical history of the grain.

—I— Magnification: (A) 500x; (B) 1,000x; (C) 20,000x; (D) 50,000x

Carbonates—Calcite

Energy Dispersive X—Ray Spectrum (EDX)

Calcite Ca CO₃

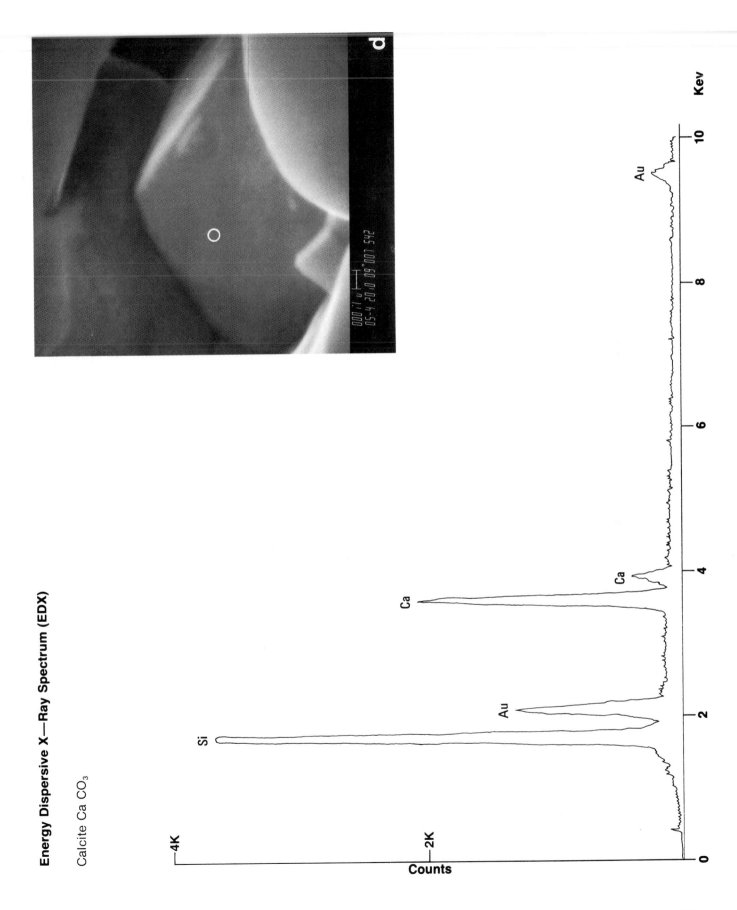

Carbonates—Calcite

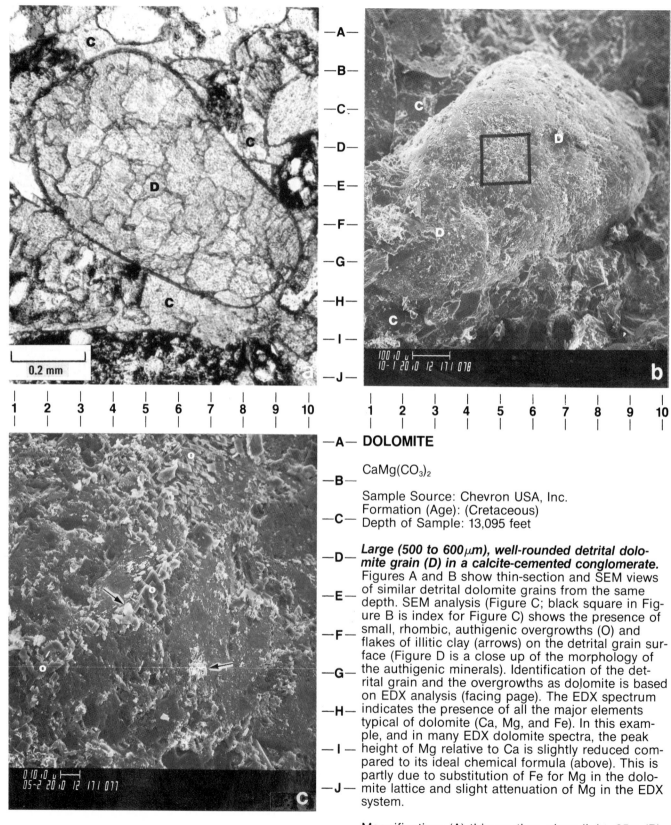

DOLOMITE

CaMg(CO$_3$)$_2$

Sample Source: Chevron USA, Inc.
Formation (Age): (Cretaceous)
Depth of Sample: 13,095 feet

Large (500 to 600 μm), well-rounded detrital dolomite grain (D) in a calcite-cemented conglomerate.
Figures A and B show thin-section and SEM views of similar detrital dolomite grains from the same depth. SEM analysis (Figure C; black square in Figure B is index for Figure C) shows the presence of small, rhombic, authigenic overgrowths (O) and flakes of illitic clay (arrows) on the detrital grain surface (Figure D is a close up of the morphology of the authigenic minerals). Identification of the detrital grain and the overgrowths as dolomite is based on EDX analysis (facing page). The EDX spectrum indicates the presence of all the major elements typical of dolomite (Ca, Mg, and Fe). In this example, and in many EDX dolomite spectra, the peak height of Mg relative to Ca is slightly reduced compared to its ideal chemical formula (above). This is partly due to substitution of Fe for Mg in the dolomite lattice and slight attenuation of Mg in the EDX system.

Magnification: (A) thin section, plane light, 25x; (B) 100x; (C) 500x; (D) 2,000x

Energy Dispersive X—Ray Spectrum (EDX)

Dolomite Ca Mg (CO₃)₂

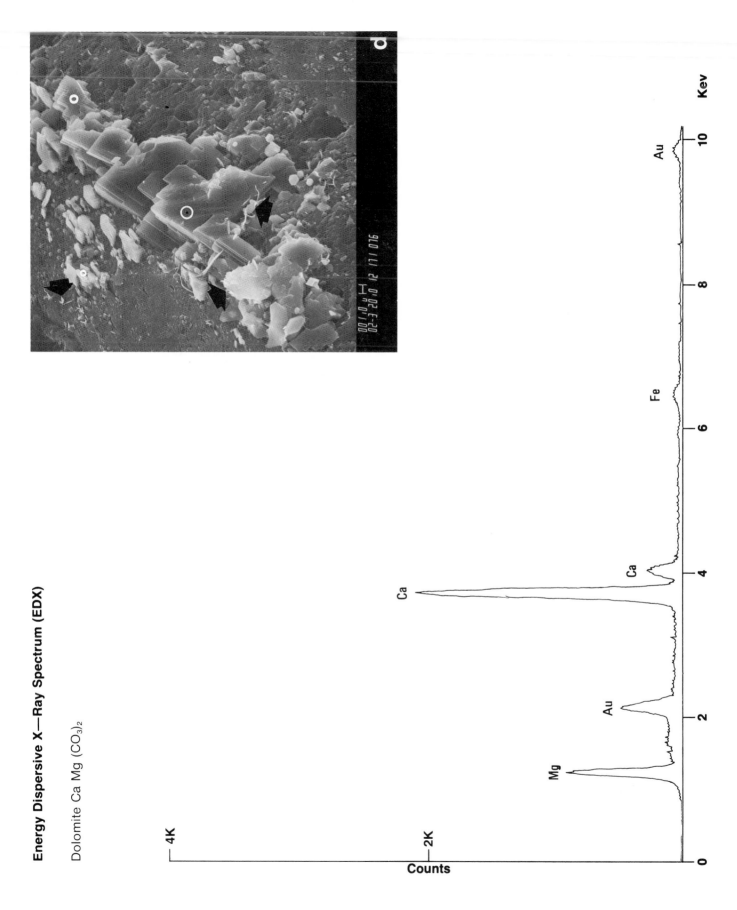

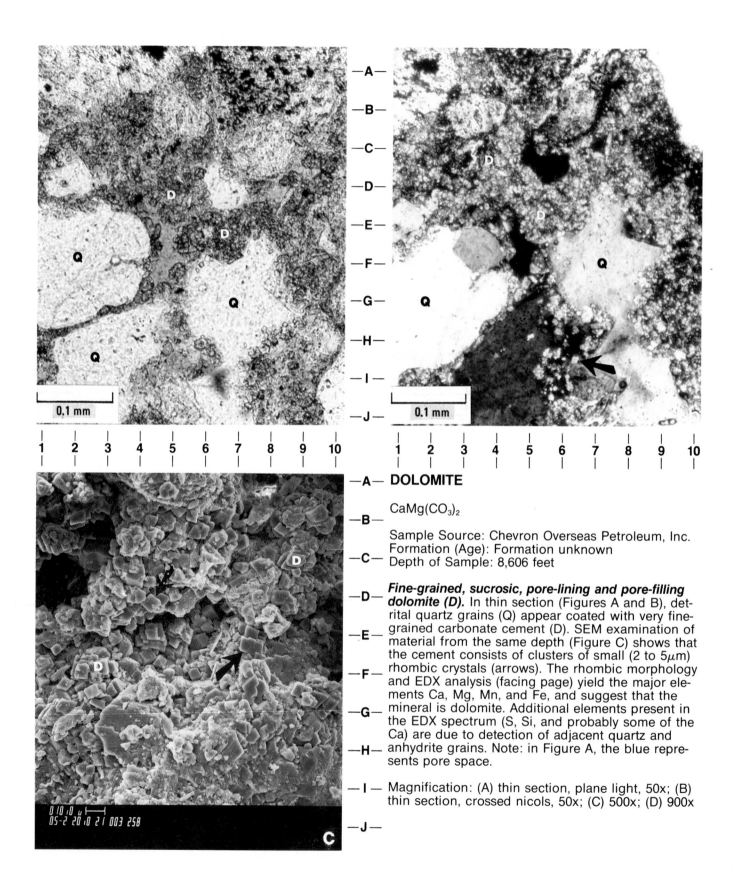

—B—

—C—

—D—

—E—

—F—

—G—

—H—

—I—

—J—

DOLOMITE

$CaMg(CO_3)_2$

Sample Source: Chevron Overseas Petroleum, Inc.
Formation (Age): Formation unknown
Depth of Sample: 8,606 feet

***Fine-grained, sucrosic, pore-lining and pore-filling
dolomite (D).*** In thin section (Figures A and B), detrital quartz grains (Q) appear coated with very fine-grained carbonate cement (D). SEM examination of material from the same depth (Figure C) shows that the cement consists of clusters of small (2 to 5μm) rhombic crystals (arrows). The rhombic morphology and EDX analysis (facing page) yield the major elements Ca, Mg, Mn, and Fe, and suggest that the mineral is dolomite. Additional elements present in the EDX spectrum (S, Si, and probably some of the Ca) are due to detection of adjacent quartz and anhydrite grains. Note: in Figure A, the blue represents pore space.

Magnification: (A) thin section, plane light, 50x; (B) thin section, crossed nicols, 50x; (C) 500x; (D) 900x

Carbonates—Dolomite

Energy Dispersive X—Ray Spectrum (EDX)

Dolomite Ca Mg (CO$_3$)$_2$

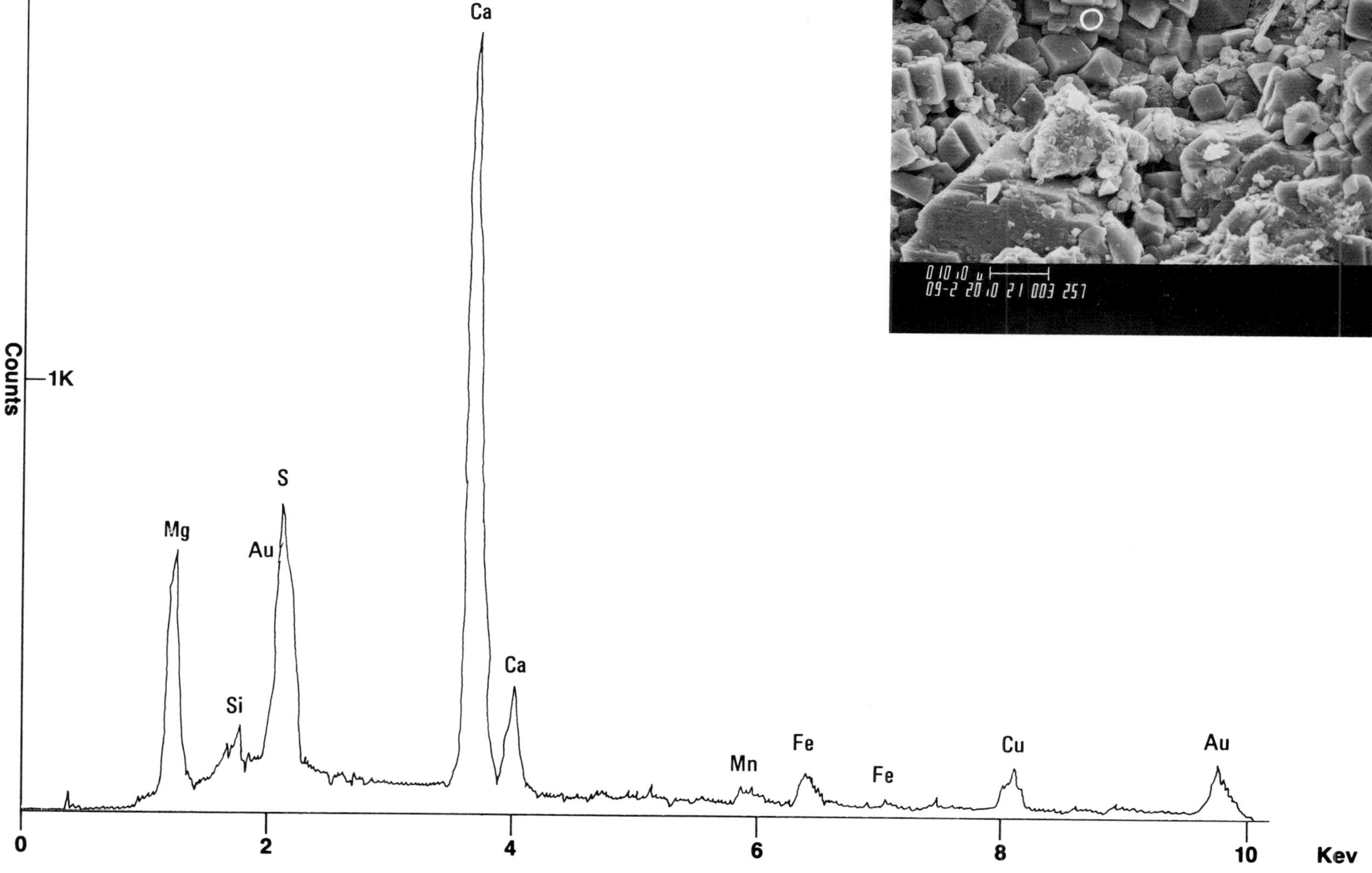

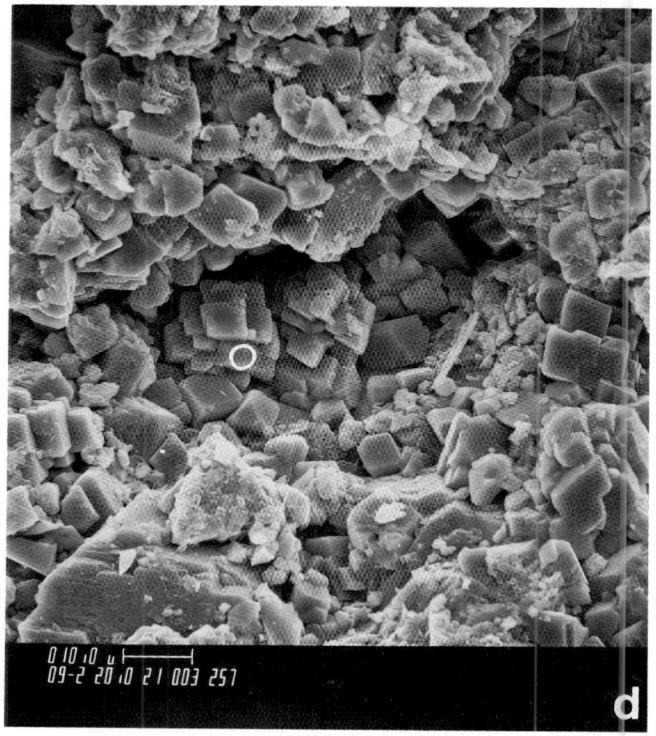

Carbonates—Dolomite

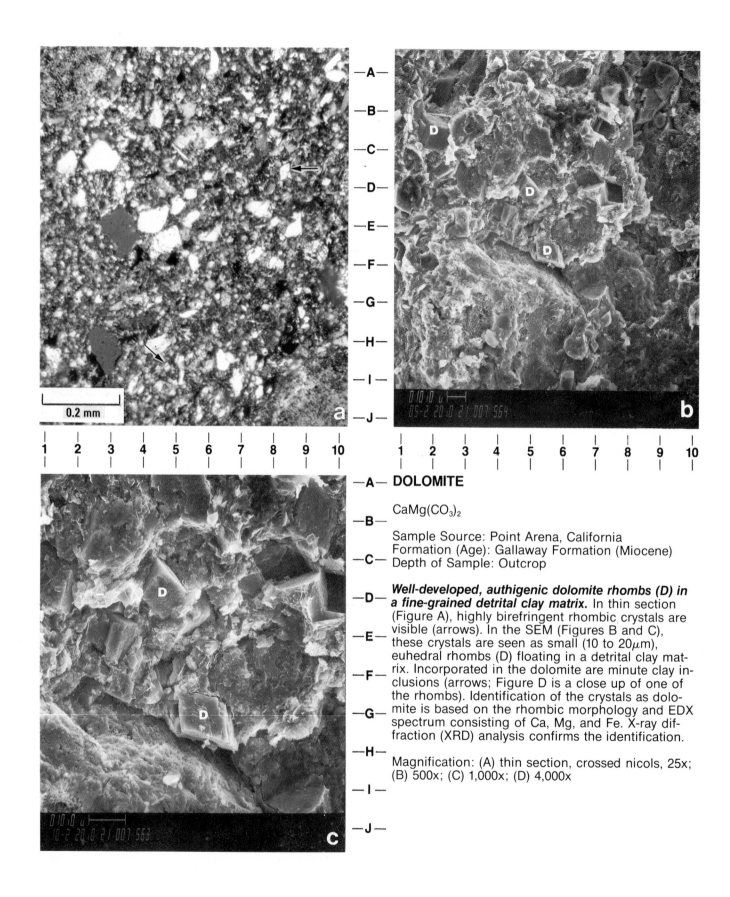

A B C D E F G H I J

DOLOMITE

$CaMg(CO_3)_2$

Sample Source: Point Arena, California
Formation (Age): Gallaway Formation (Miocene)
Depth of Sample: Outcrop

Well-developed, authigenic dolomite rhombs (D) in a fine-grained detrital clay matrix. In thin section (Figure A), highly birefringent rhombic crystals are visible (arrows). In the SEM (Figures B and C), these crystals are seen as small (10 to 20μm), euhedral rhombs (D) floating in a detrital clay matrix. Incorporated in the dolomite are minute clay inclusions (arrows; Figure D is a close up of one of the rhombs). Identification of the crystals as dolomite is based on the rhombic morphology and EDX spectrum consisting of Ca, Mg, and Fe. X-ray diffraction (XRD) analysis confirms the identification.

Magnification: (A) thin section, crossed nicols, 25x; (B) 500x; (C) 1,000x; (D) 4,000x

Carbonates—Dolomite

Energy Dispersive X—Ray Spectrum (EDX)

Dolomite Ca Mg (CO$_3$)$_2$

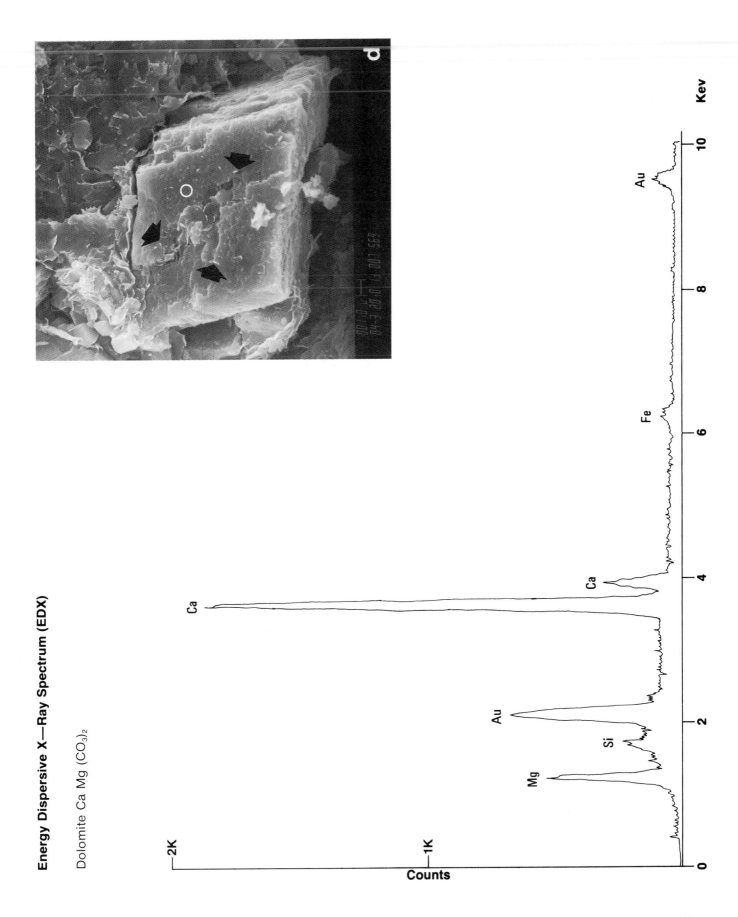

Carbonates—Dolomite

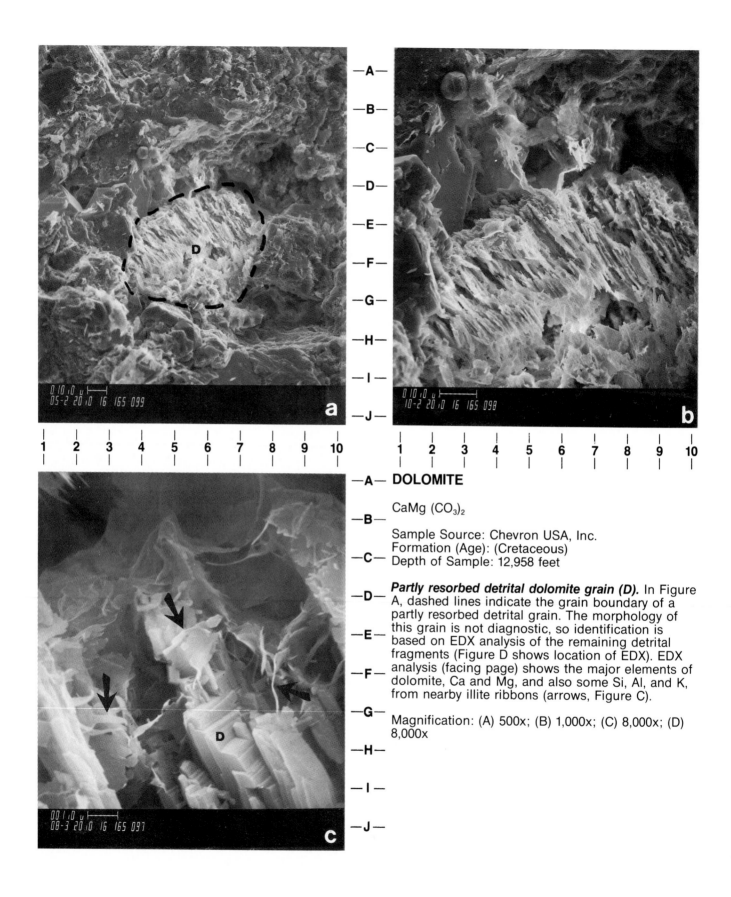

DOLOMITE

CaMg (CO$_3$)$_2$

Sample Source: Chevron USA, Inc.
Formation (Age): (Cretaceous)
Depth of Sample: 12,958 feet

Partly resorbed detrital dolomite grain (D). In Figure
A, dashed lines indicate the grain boundary of a
partly resorbed detrital grain. The morphology of
this grain is not diagnostic, so identification is
based on EDX analysis of the remaining detrital
fragments (Figure D shows location of EDX). EDX
analysis (facing page) shows the major elements of
dolomite, Ca and Mg, and also some Si, Al, and K,
from nearby illite ribbons (arrows, Figure C).

Magnification: (A) 500x; (B) 1,000x; (C) 8,000x; (D)
8,000x

Carbonates-Dolomite

Energy Dispersive X—Ray Spectrum (EDX)

Dolomite Ca Mg (CO$_3$)$_2$

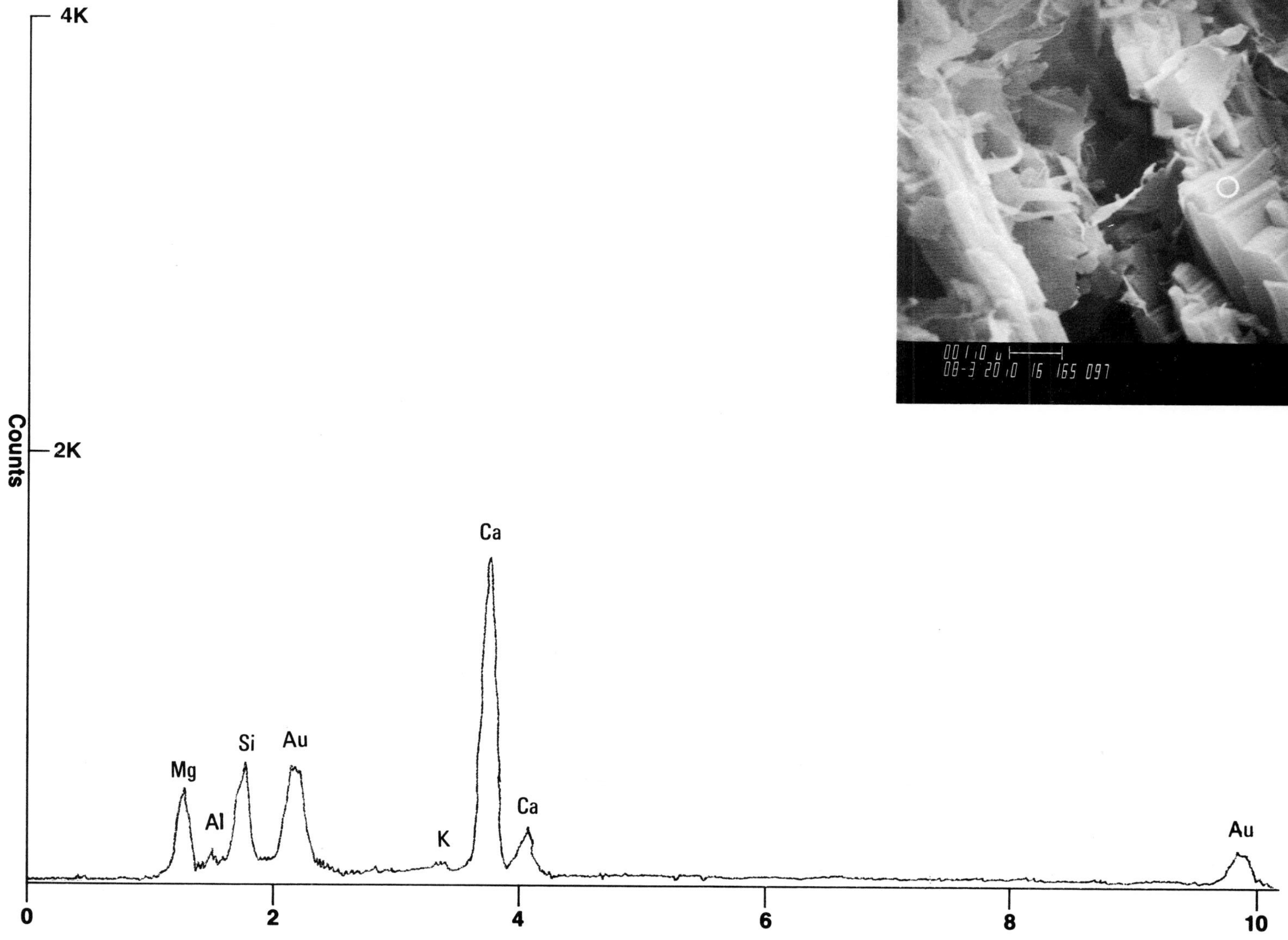

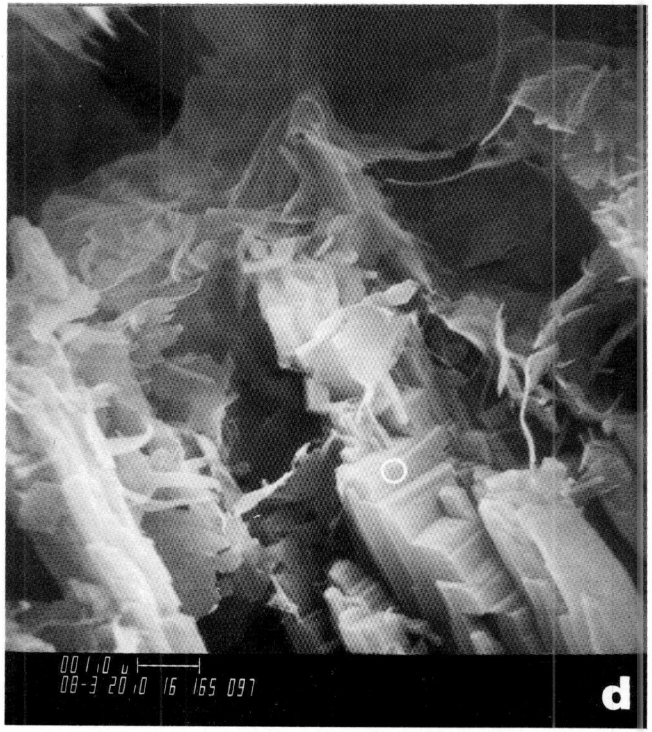

Carbonates-Dolomite

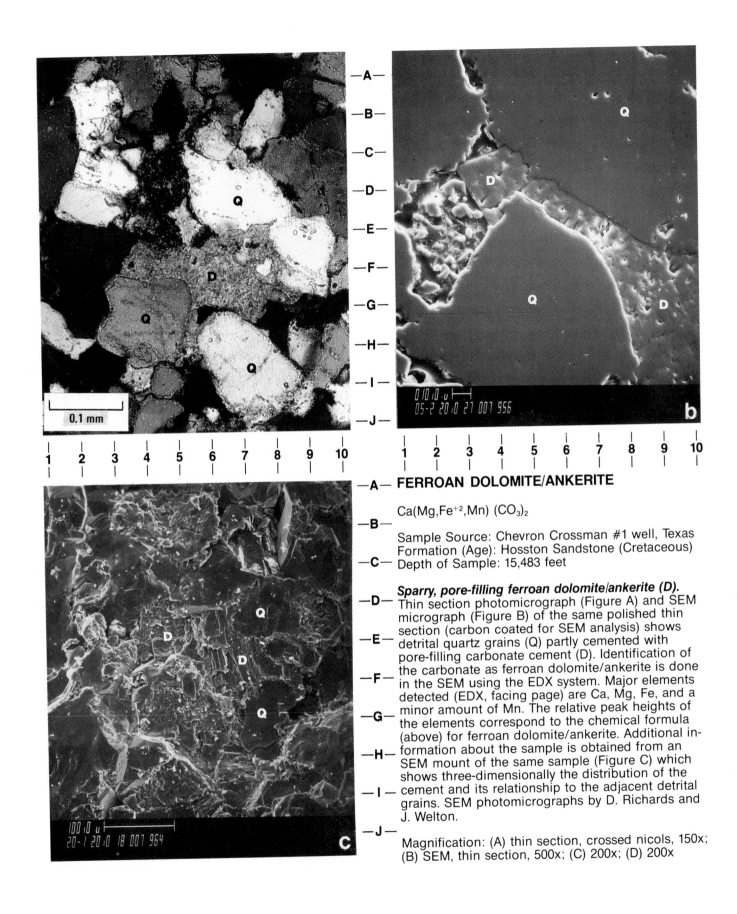

0.1 mm

1 2 3 4 5 6 7 8 9 10

FERROAN DOLOMITE/ANKERITE

$Ca(Mg,Fe^{+2},Mn)(CO_3)_2$

Sample Source: Chevron Crossman #1 well, Texas
Formation (Age): Hosston Sandstone (Cretaceous)
Depth of Sample: 15,483 feet

Sparry, pore-filling ferroan dolomite/ankerite (D).
Thin section photomicrograph (Figure A) and SEM micrograph (Figure B) of the same polished thin section (carbon coated for SEM analysis) shows detrital quartz grains (Q) partly cemented with pore-filling carbonate cement (D). Identification of the carbonate as ferroan dolomite/ankerite is done in the SEM using the EDX system. Major elements detected (EDX, facing page) are Ca, Mg, Fe, and a minor amount of Mn. The relative peak heights of the elements correspond to the chemical formula (above) for ferroan dolomite/ankerite. Additional information about the sample is obtained from an SEM mount of the same sample (Figure C) which shows three-dimensionally the distribution of the cement and its relationship to the adjacent detrital grains. SEM photomicrographs by D. Richards and J. Welton.

Magnification: (A) thin section, crossed nicols, 150x; (B) SEM, thin section, 500x; (C) 200x; (D) 200x

Carbonates—Ankerite

Energy Dispersive X—Ray Spectrum (EDX)

Ferroan Dolomite/Ankerite Ca (Mg, Fe^{+2}, Mn) $(CO_3)_2$

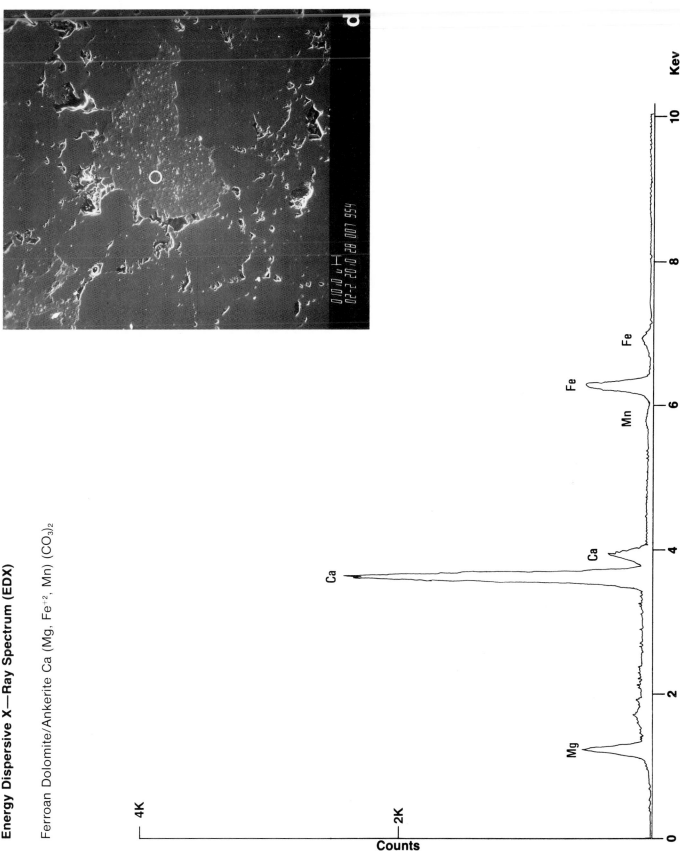

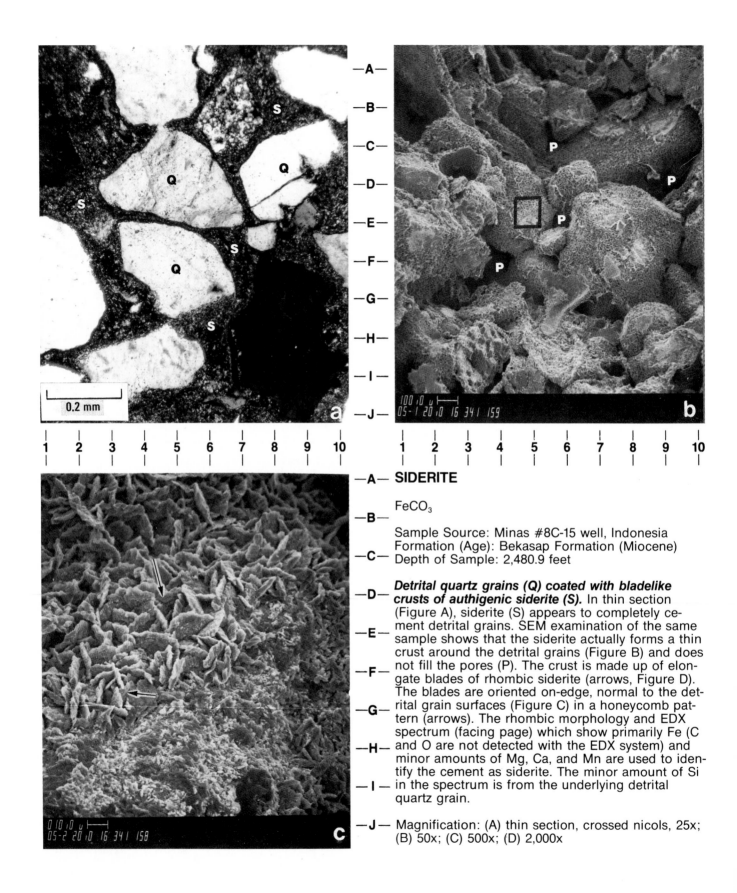

SIDERITE

FeCO$_3$

Sample Source: Minas #8C-15 well, Indonesia
Formation (Age): Bekasap Formation (Miocene)
Depth of Sample: 2,480.9 feet

Detrital quartz grains (Q) coated with bladelike crusts of authigenic siderite (S). In thin section (Figure A), siderite (S) appears to completely cement detrital grains. SEM examination of the same sample shows that the siderite actually forms a thin crust around the detrital grains (Figure B) and does not fill the pores (P). The crust is made up of elongate blades of rhombic siderite (arrows, Figure D). The blades are oriented on-edge, normal to the detrital grain surfaces (Figure C) in a honeycomb pattern (arrows). The rhombic morphology and EDX spectrum (facing page) which show primarily Fe (C and O are not detected with the EDX system) and minor amounts of Mg, Ca, and Mn are used to identify the cement as siderite. The minor amount of Si in the spectrum is from the underlying detrital quartz grain.

Magnification: (A) thin section, crossed nicols, 25x; (B) 50x; (C) 500x; (D) 2,000x

Energy Dispersive X—Ray Spectrum (EDX)

Siderite Fe CO$_3$

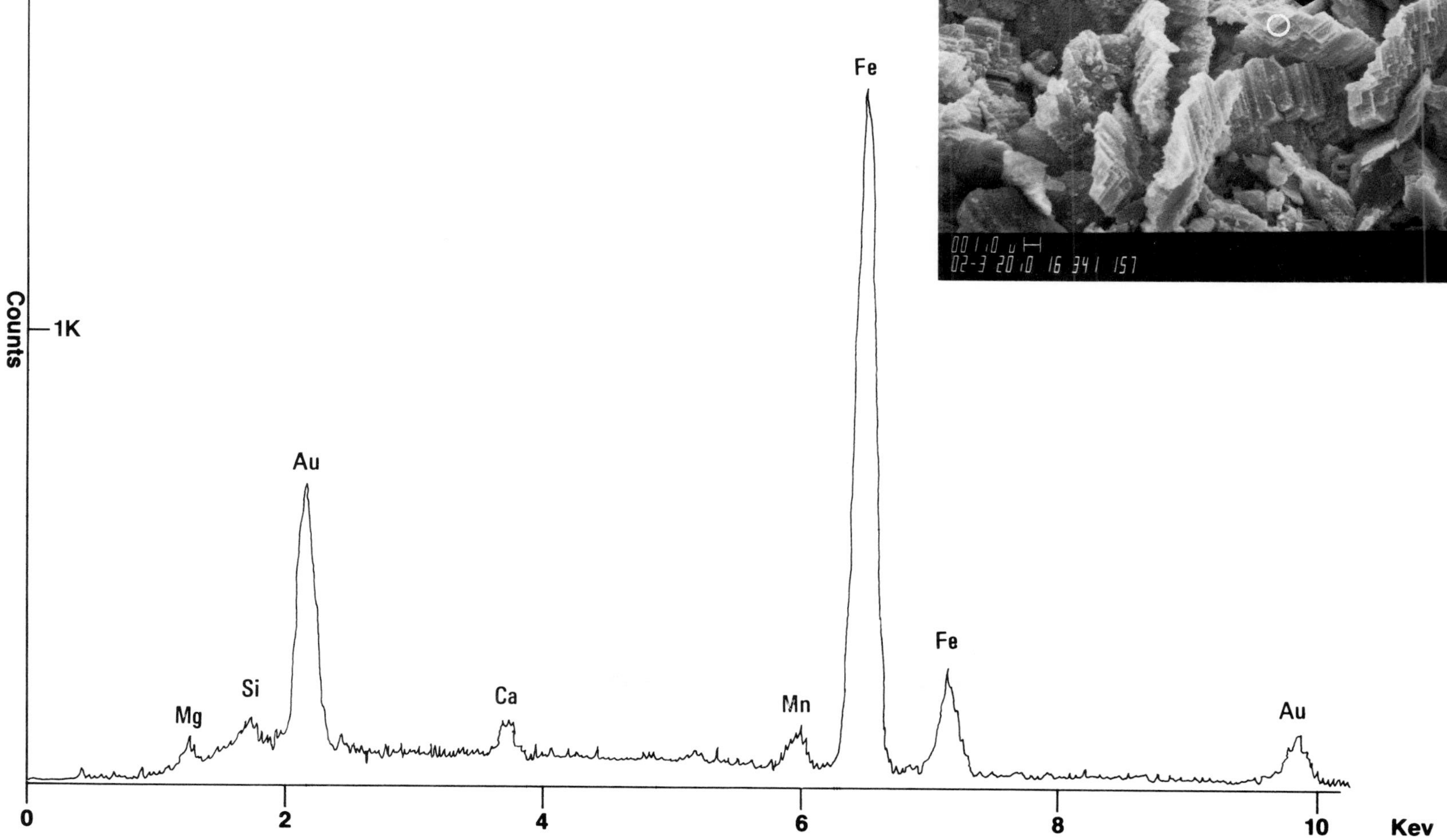

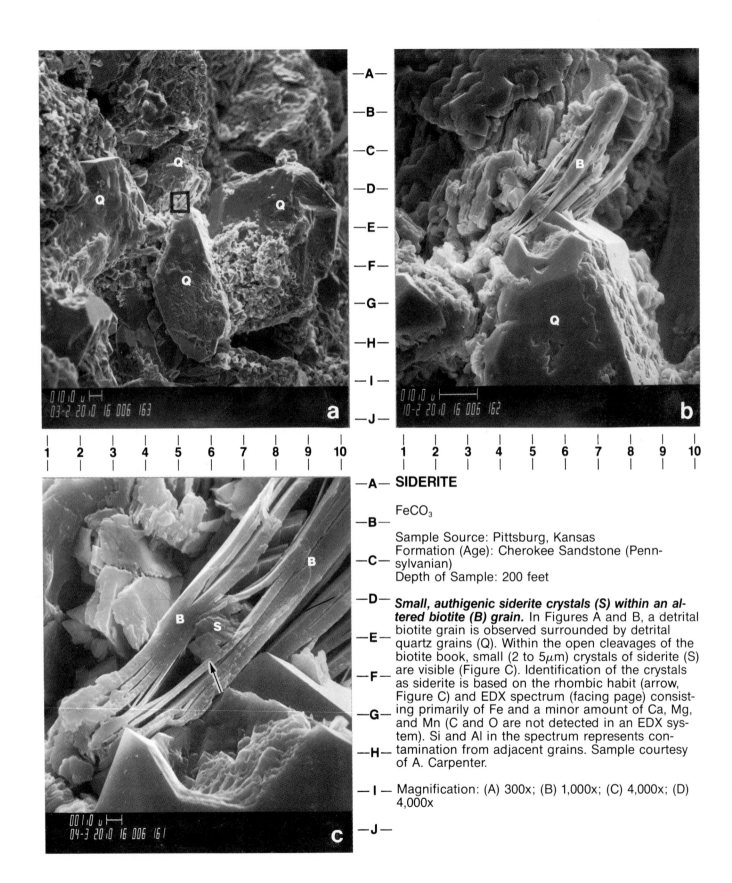

SIDERITE

FeCO$_3$

Sample Source: Pittsburg, Kansas
Formation (Age): Cherokee Sandstone (Pennsylvanian)
Depth of Sample: 200 feet

Small, authigenic siderite crystals (S) within an altered biotite (B) grain. In Figures A and B, a detrital biotite grain is observed surrounded by detrital quartz grains (Q). Within the open cleavages of the biotite book, small (2 to 5μm) crystals of siderite (S) are visible (Figure C). Identification of the crystals as siderite is based on the rhombic habit (arrow, Figure C) and EDX spectrum (facing page) consisting primarily of Fe and a minor amount of Ca, Mg, and Mn (C and O are not detected in an EDX system). Si and Al in the spectrum represents contamination from adjacent grains. Sample courtesy of A. Carpenter.

Magnification: (A) 300x; (B) 1,000x; (C) 4,000x; (D) 4,000x

Energy Dispersive X—Ray Spectrum (EDX)

Siderite Fe CO₃

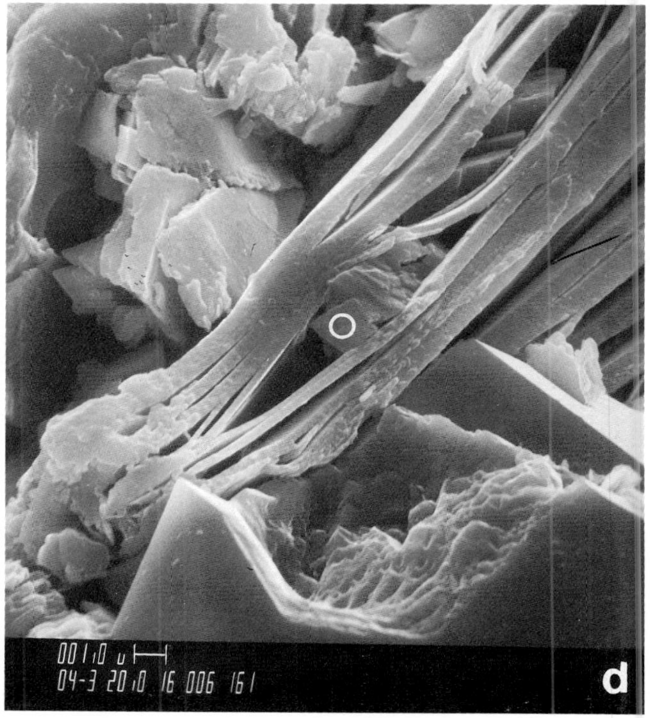

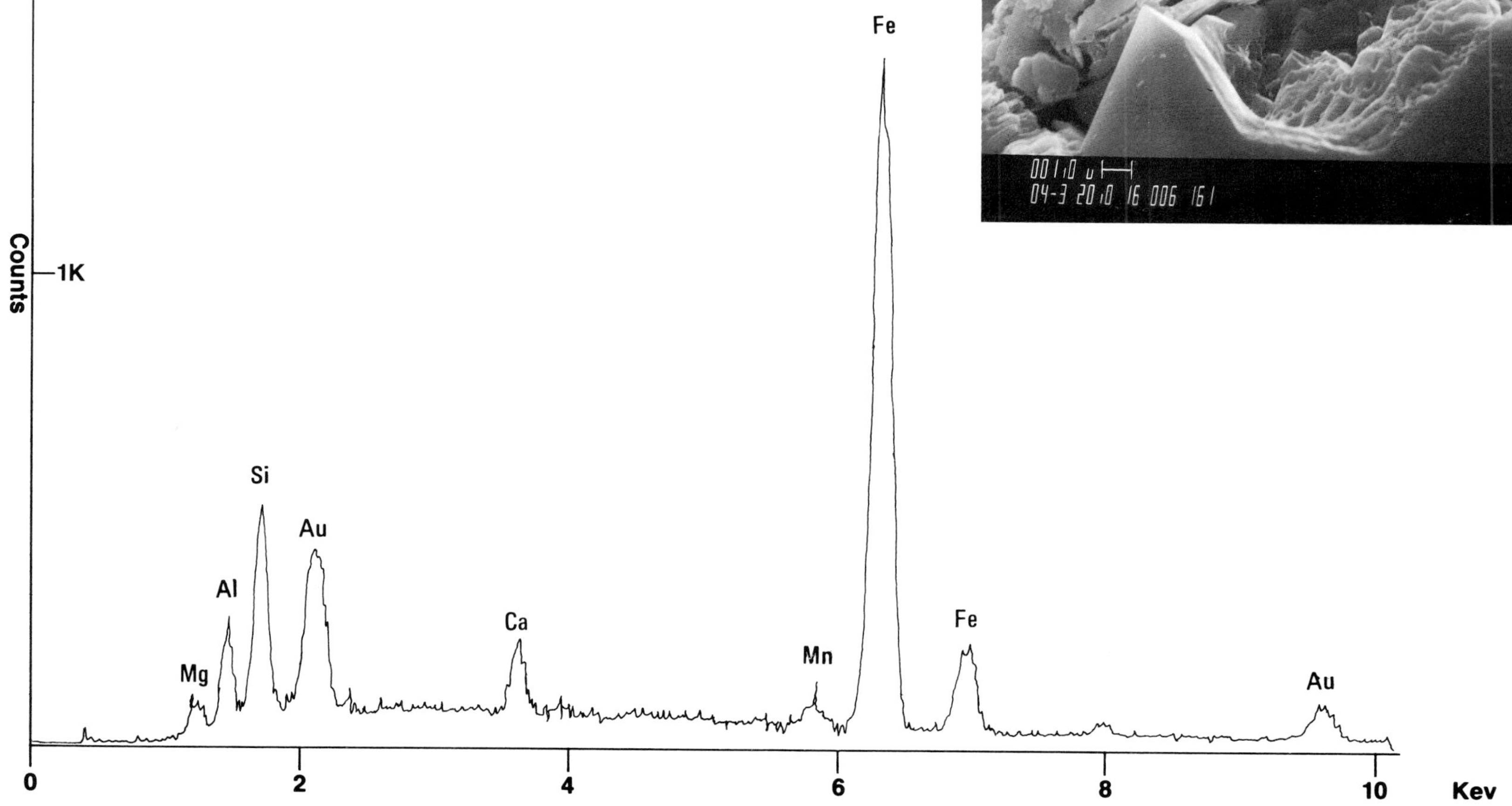

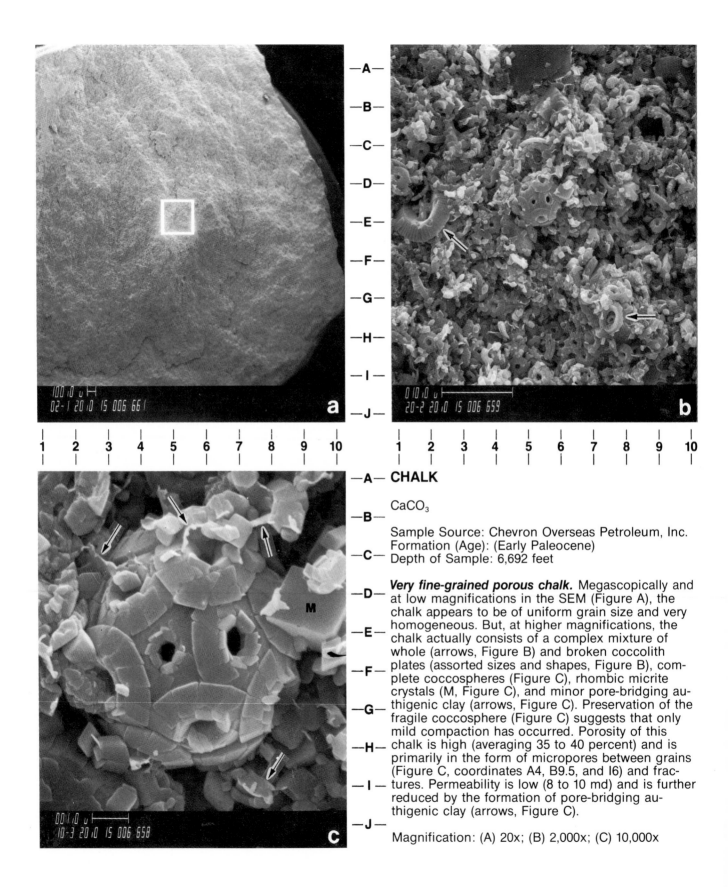

—A— CHALK

CaCO₃

Sample Source: Chevron Overseas Petroleum, Inc.
Formation (Age): (Early Paleocene)
Depth of Sample: 6,692 feet

Very fine-grained porous chalk. Megascopically and at low magnifications in the SEM (Figure A), the chalk appears to be of uniform grain size and very homogeneous. But, at higher magnifications, the chalk actually consists of a complex mixture of whole (arrows, Figure B) and broken coccolith plates (assorted sizes and shapes, Figure B), complete coccospheres (Figure C), rhombic micrite crystals (M, Figure C), and minor pore-bridging authigenic clay (arrows, Figure C). Preservation of the fragile coccosphere (Figure C) suggests that only mild compaction has occurred. Porosity of this chalk is high (averaging 35 to 40 percent) and is primarily in the form of micropores between grains (Figure C, coordinates A4, B9.5, and I6) and fractures. Permeability is low (8 to 10 md) and is further reduced by the formation of pore-bridging authigenic clay (arrows, Figure C).

Magnification: (A) 20x; (B) 2,000x; (C) 10,000x

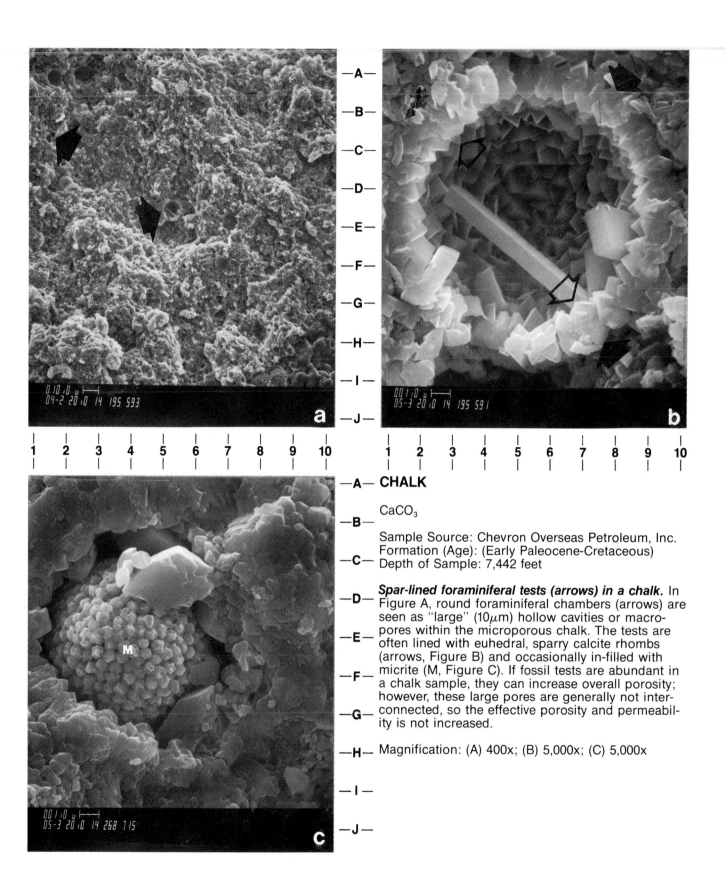

—A— **CHALK**

—B— CaCO₃

—C— Sample Source: Chevron Overseas Petroleum, Inc.
Formation (Age): (Early Paleocene-Cretaceous)
Depth of Sample: 7,442 feet

—D— ***Spar-lined foraminiferal tests (arrows) in a chalk.*** In
Figure A, round foraminiferal chambers (arrows) are
seen as "large" (10μm) hollow cavities or macro-
—E— pores within the microporous chalk. The tests are
often lined with euhedral, sparry calcite rhombs
(arrows, Figure B) and occasionally in-filled with
—F— micrite (M, Figure C). If fossil tests are abundant in
a chalk sample, they can increase overall porosity;
however, these large pores are generally not inter-
—G— connected, so the effective porosity and permeability is not increased.

—H— Magnification: (A) 400x; (B) 5,000x; (C) 5,000x

APATITE

$Ca_5(PO_4)_3$ (OH,F,Cl)

Sample Source: Point Arena, California
Formation (Age): Gallaway Formation (Miocene)
Depth of Sample: Outcrop

Small (20 to 30 μm), equant, detrital apatite crystal (A) in an argillaceous lithic arkose. Identification of this crystal as apatite is based on EDX analysis (facing page) showing the major elements of apatite, Ca, P, and Cl. Minor amounts of Si, Al, Mg, and Fe in the EDX spectrum are probably contaminants from surrounding clay minerals.

Magnification: (A) 200x; (B) 500x; (C) 1,000x; (D) 2,000x

Energy Dispersive X—Ray Spectrum (EDX)

Apatite $Ca_5 (PO_4)_3 (OH, F, Cl)$

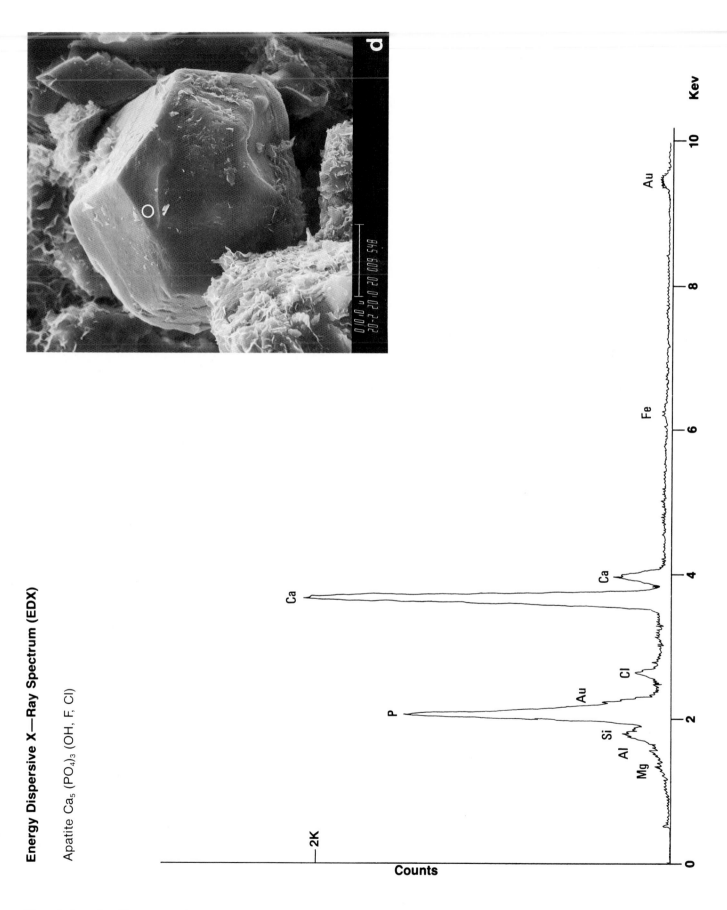

Phosphates—Apatite

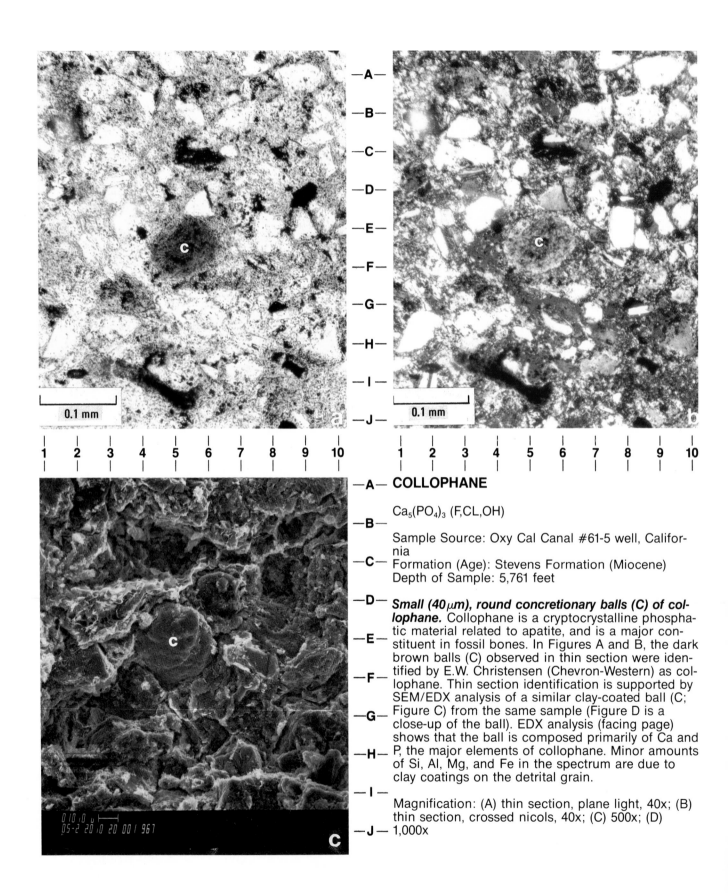

COLLOPHANE

$Ca_5(PO_4)_3$ (F,CL,OH)

Sample Source: Oxy Cal Canal #61-5 well, California

Formation (Age): Stevens Formation (Miocene)

Depth of Sample: 5,761 feet

Small (40 μm), round concretionary balls (C) of collophane. Collophane is a cryptocrystalline phosphatic material related to apatite, and is a major constituent in fossil bones. In Figures A and B, the dark brown balls (C) observed in thin section were identified by E.W. Christensen (Chevron-Western) as collophane. Thin section identification is supported by SEM/EDX analysis of a similar clay-coated ball (C; Figure C) from the same sample (Figure D is a close-up of the ball). EDX analysis (facing page) shows that the ball is composed primarily of Ca and P, the major elements of collophane. Minor amounts of Si, Al, Mg, and Fe in the spectrum are due to clay coatings on the detrital grain.

Magnification: (A) thin section, plane light, 40x; (B) thin section, crossed nicols, 40x; (C) 500x; (D) 1,000x

Phosphates—Collophane

Energy Dispersive X—Ray Spectrum (EDX)

Collophane Ca$_5$ (PO$_4$)$_3$ (F, Cl, OH)

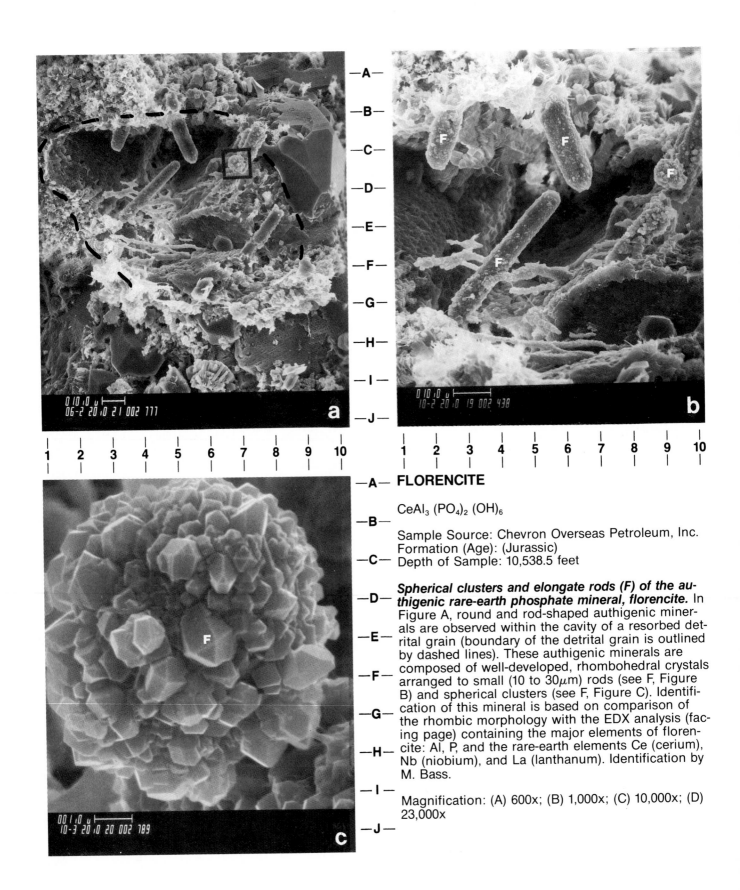

FLORENCITE

$CeAl_3 (PO_4)_2 (OH)_6$

Sample Source: Chevron Overseas Petroleum, Inc.
Formation (Age): (Jurassic)
Depth of Sample: 10,538.5 feet

Spherical clusters and elongate rods (F) of the authigenic rare-earth phosphate mineral, florencite. In Figure A, round and rod-shaped authigenic minerals are observed within the cavity of a resorbed detrital grain (boundary of the detrital grain is outlined by dashed lines). These authigenic minerals are composed of well-developed, rhombohedral crystals arranged to small (10 to 30μm) rods (see F, Figure B) and spherical clusters (see F, Figure C). Identification of this mineral is based on comparison of the rhombic morphology with the EDX analysis (facing page) containing the major elements of florencite: Al, P, and the rare-earth elements Ce (cerium), Nb (niobium), and La (lanthanum). Identification by M. Bass.

Magnification: (A) 600x; (B) 1,000x; (C) 10,000x; (D) 23,000x

Energy Dispersive X—Ray Spectrum (EDX)

Florencite Ce Al$_3$ (PO$_4$)$_2$ (OH)$_6$

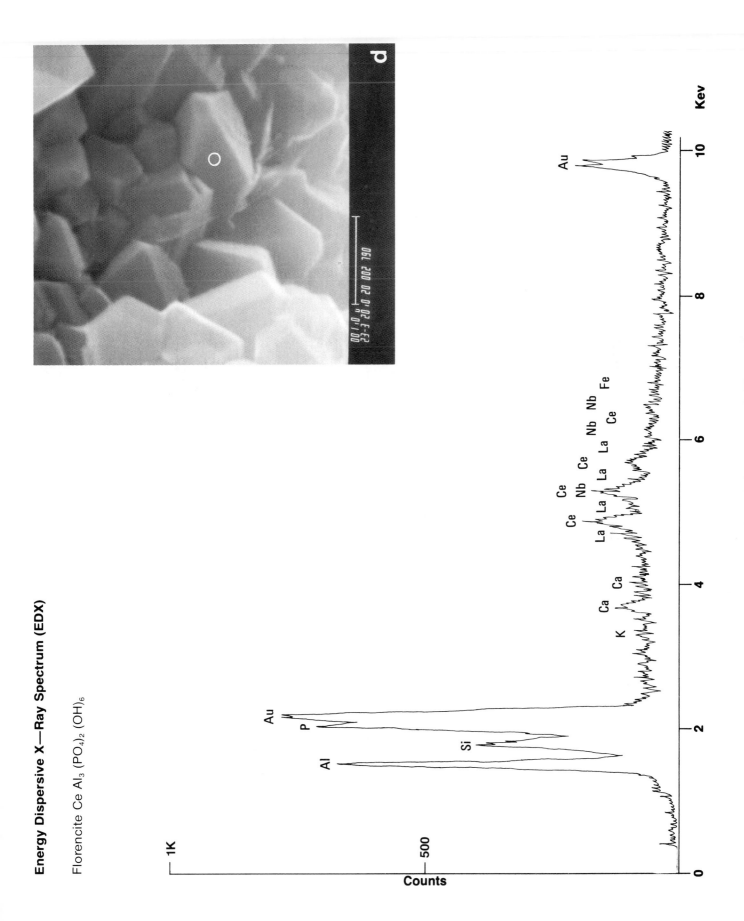

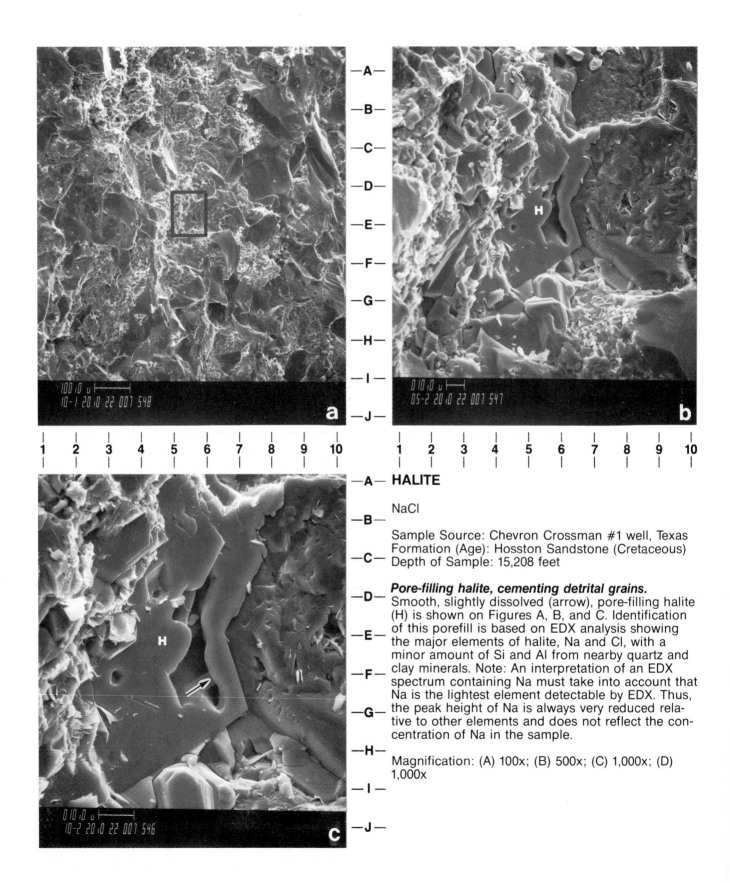

| 1 | 2 | 3 | 4 | 5 | 6 | 7 | 8 | 9 | 10 |

HALITE

NaCl

Sample Source: Chevron Crossman #1 well, Texas
Formation (Age): Hosston Sandstone (Cretaceous)
Depth of Sample: 15,208 feet

Pore-filling halite, cementing detrital grains.
Smooth, slightly dissolved (arrow), pore-filling halite
(H) is shown on Figures A, B, and C. Identification
of this porefill is based on EDX analysis showing
the major elements of halite, Na and Cl, with a
minor amount of Si and Al from nearby quartz and
clay minerals. Note: An interpretation of an EDX
spectrum containing Na must take into account that
Na is the lightest element detectable by EDX. Thus,
the peak height of Na is always very reduced relative to other elements and does not reflect the concentration of Na in the sample.

Magnification: (A) 100x; (B) 500x; (C) 1,000x; (D)
1,000x

Energy Dispersive X—Ray Spectrum (EDX)

Halite Na Cl

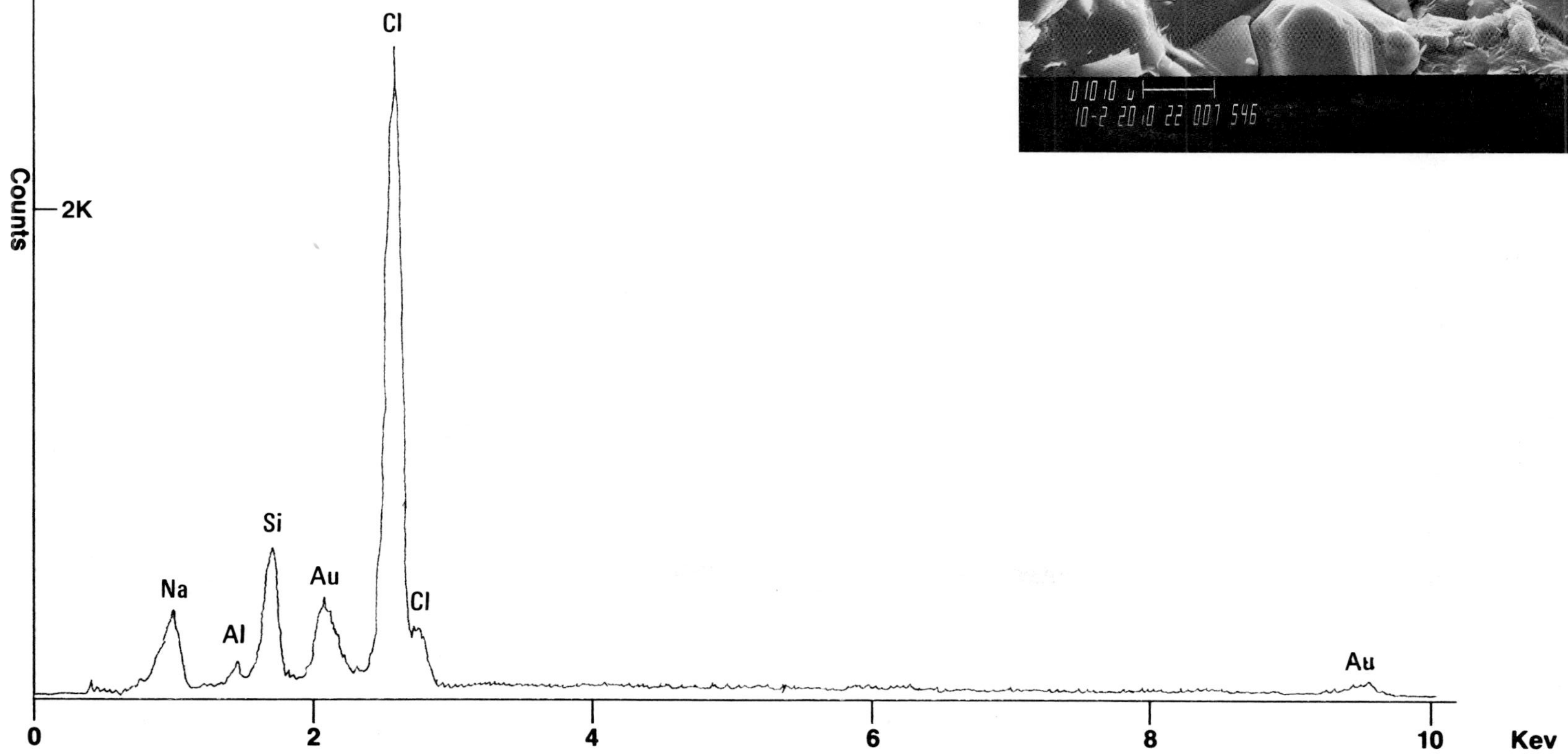

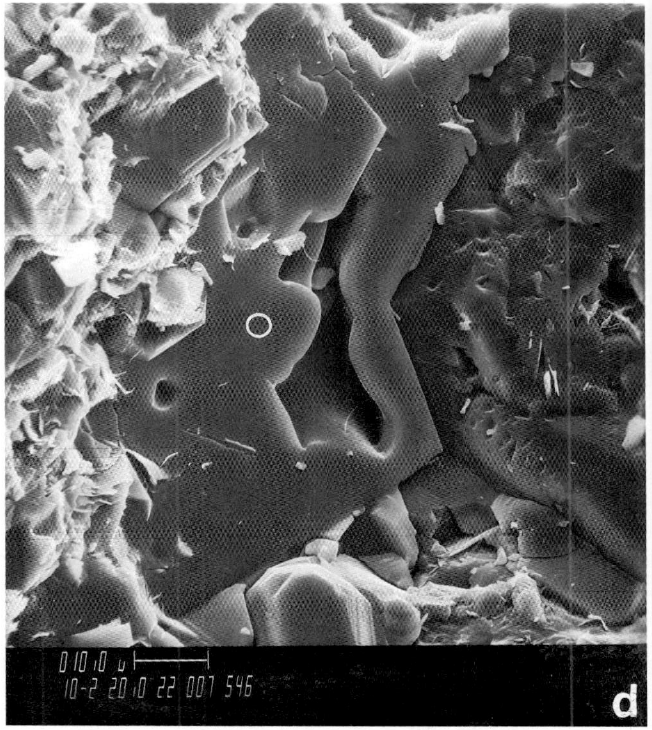

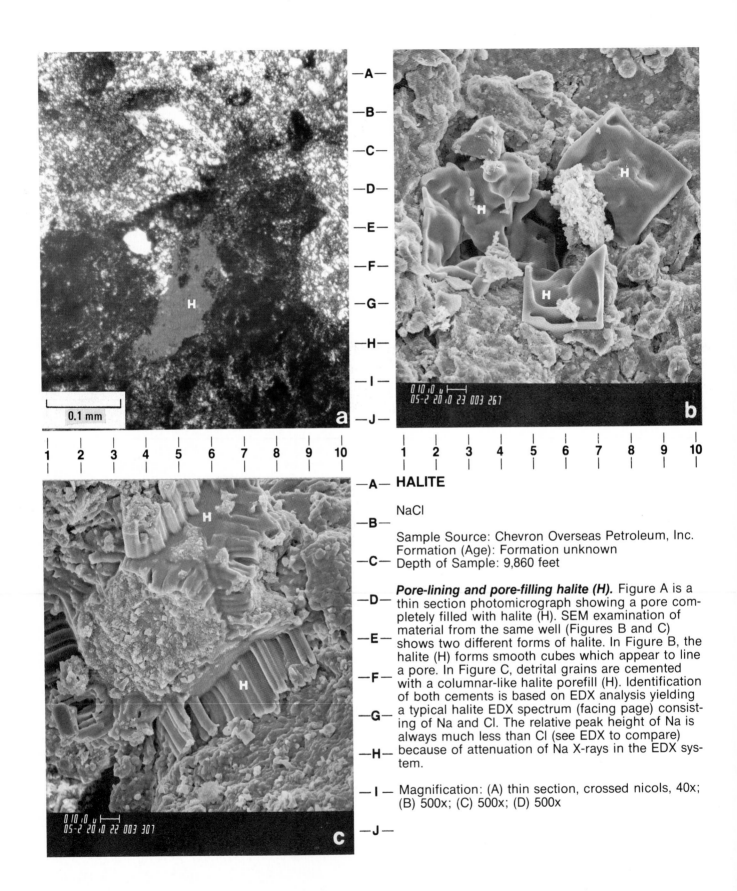

HALITE

NaCl

Sample Source: Chevron Overseas Petroleum, Inc.
Formation (Age): Formation unknown
Depth of Sample: 9,860 feet

Pore-lining and pore-filling halite (H). Figure A is a thin section photomicrograph showing a pore completely filled with halite (H). SEM examination of material from the same well (Figures B and C) shows two different forms of halite. In Figure B, the halite (H) forms smooth cubes which appear to line a pore. In Figure C, detrital grains are cemented with a columnar-like halite porefill (H). Identification of both cements is based on EDX analysis yielding a typical halite EDX spectrum (facing page) consisting of Na and Cl. The relative peak height of Na is always much less than Cl (see EDX to compare) because of attenuation of Na X-rays in the EDX system.

Magnification: (A) thin section, crossed nicols, 40x; (B) 500x; (C) 500x; (D) 500x

Halides—Halite

Energy Dispersive X—Ray Spectrum (EDX)

Halite Na Cl

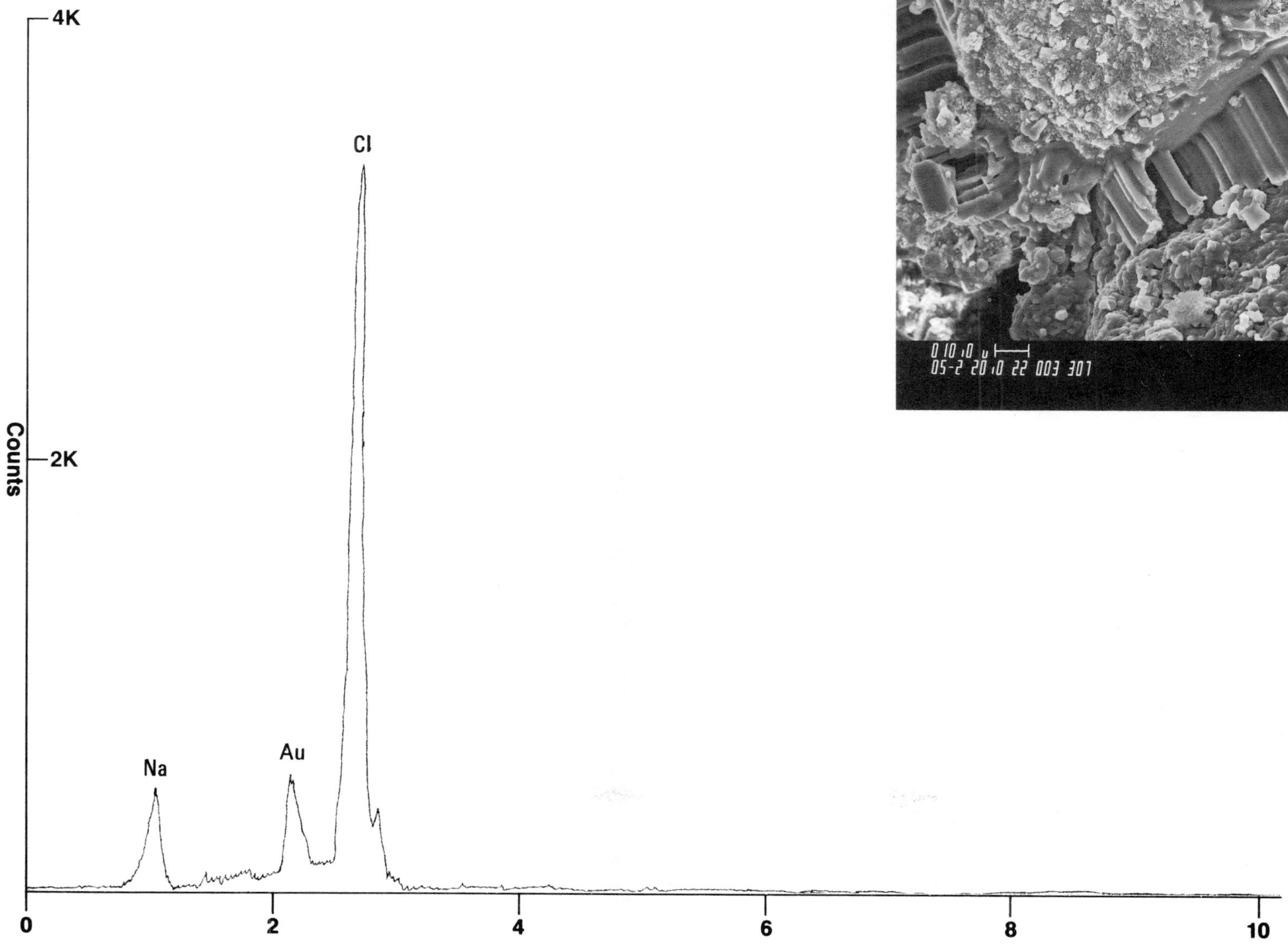

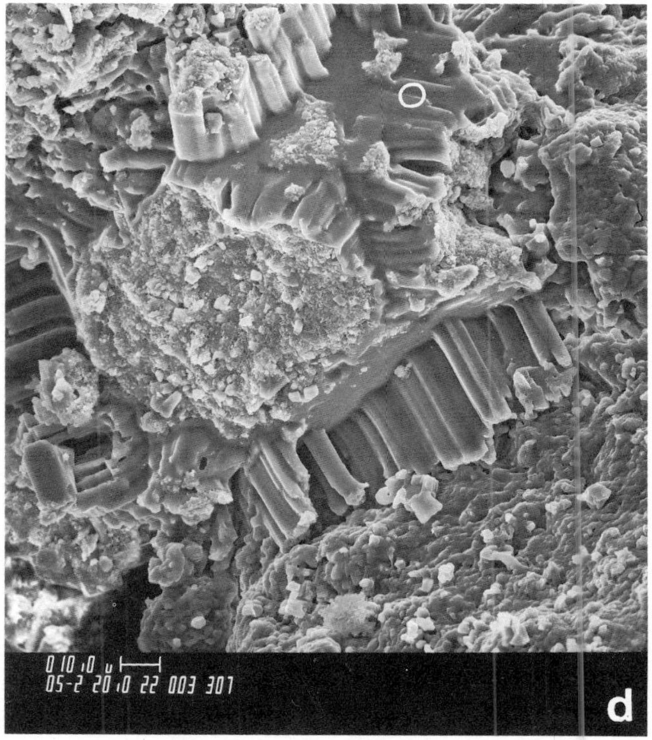

Cl

Au

Na

Counts

4K

2K

Kev

0 2 4 6 8 10

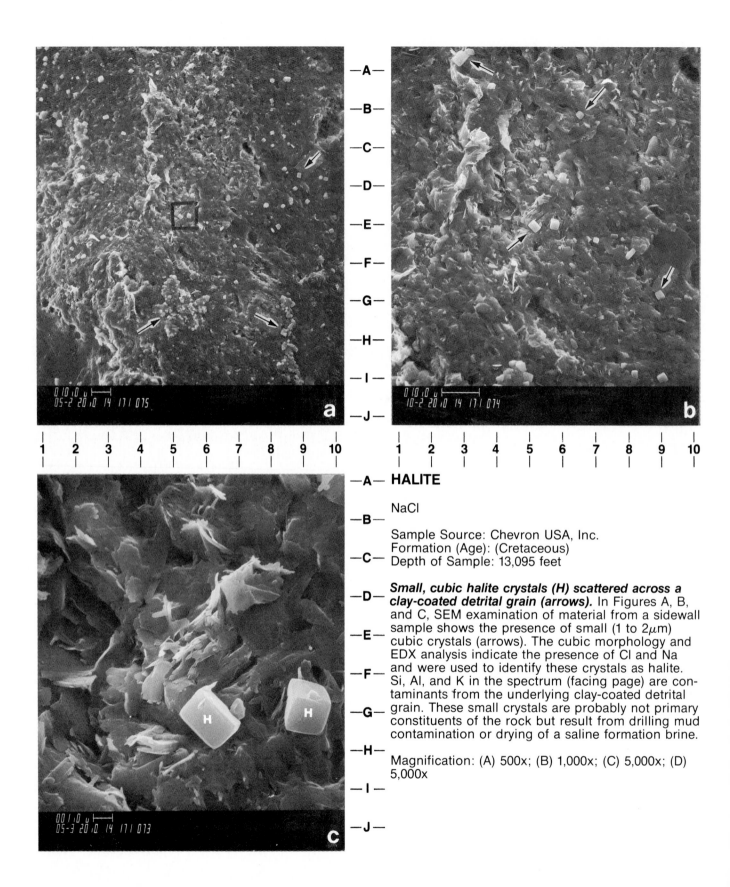

HALITE

NaCl

Sample Source: Chevron USA, Inc.
Formation (Age): (Cretaceous)
Depth of Sample: 13,095 feet

Small, cubic halite crystals (H) scattered across a clay-coated detrital grain (arrows). In Figures A, B, and C, SEM examination of material from a sidewall sample shows the presence of small (1 to 2μm) cubic crystals (arrows). The cubic morphology and EDX analysis indicate the presence of Cl and Na and were used to identify these crystals as halite. Si, Al, and K in the spectrum (facing page) are contaminants from the underlying clay-coated detrital grain. These small crystals are probably not primary constituents of the rock but result from drilling mud contamination or drying of a saline formation brine.

Magnification: (A) 500x; (B) 1,000x; (C) 5,000x; (D) 5,000x

Energy Dispersive X—Ray Spectrum (EDX)

Halite Na Cl

PYRITE

FeS$_2$

Sample Source: Chevron USA, Inc.
Formation (Age): (Cretaceous)
Depth of Sample: 12,958 feet

Pyritohedron (P) on a clay-coated detrital quartz grain with authigenic quartz overgrowths (O). In Figure B, a well-developed euhedral, authigenic pyrite crystal (P) is shown (square in Figure A is index for Figure B pyritohedron). Enlargement of the surface of the pyrite crystal (Figure C) shows a thin coating of small rods (arrows) and pseudohexagonal clay platelets (C). These rods and platelets are too small and thin to be analyzed with the EDX system. Based on morphology, the rods are probably a fibrous zeolite mineral and the clay platelets are kaolinite.

Magnification: (A) 500x; (B) 5,000x; (C) 10,000x

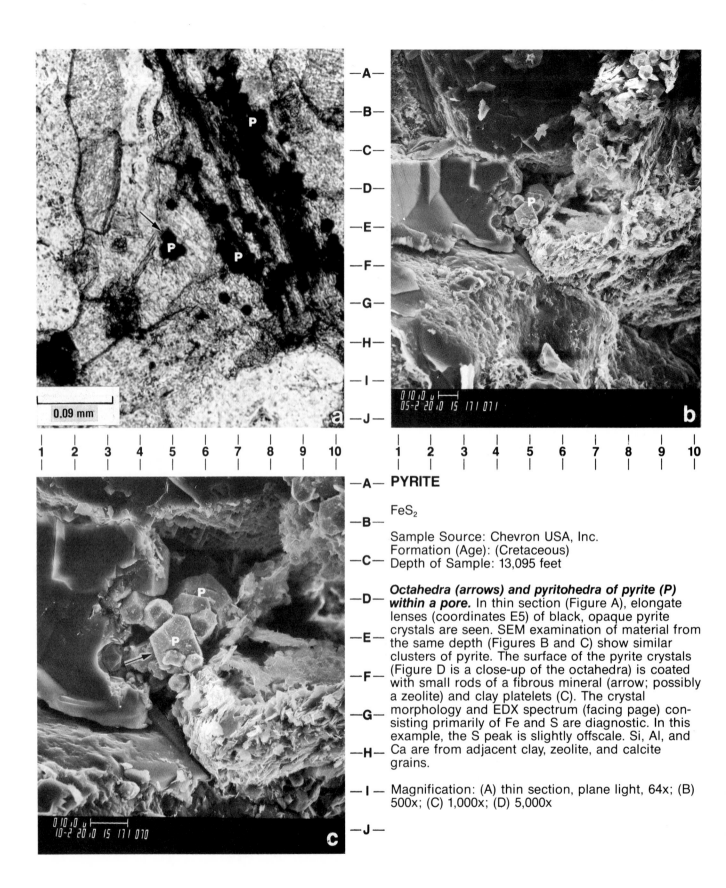

PYRITE

FeS$_2$

Sample Source: Chevron USA, Inc.
Formation (Age): (Cretaceous)
Depth of Sample: 13,095 feet

Octahedra (arrows) and pyritohedra of pyrite (P) within a pore. In thin section (Figure A), elongate lenses (coordinates E5) of black, opaque pyrite crystals are seen. SEM examination of material from the same depth (Figures B and C) show similar clusters of pyrite. The surface of the pyrite crystals (Figure D is a close-up of the octahedra) is coated with small rods of a fibrous mineral (arrow; possibly a zeolite) and clay platelets (C). The crystal morphology and EDX spectrum (facing page) consisting primarily of Fe and S are diagnostic. In this example, the S peak is slightly offscale. Si, Al, and Ca are from adjacent clay, zeolite, and calcite grains.

Magnification: (A) thin section, plane light, 64x; (B) 500x; (C) 1,000x; (D) 5,000x

Energy Dispersive X—Ray Spectrum (EDX)

Pyrite Fe S$_2$

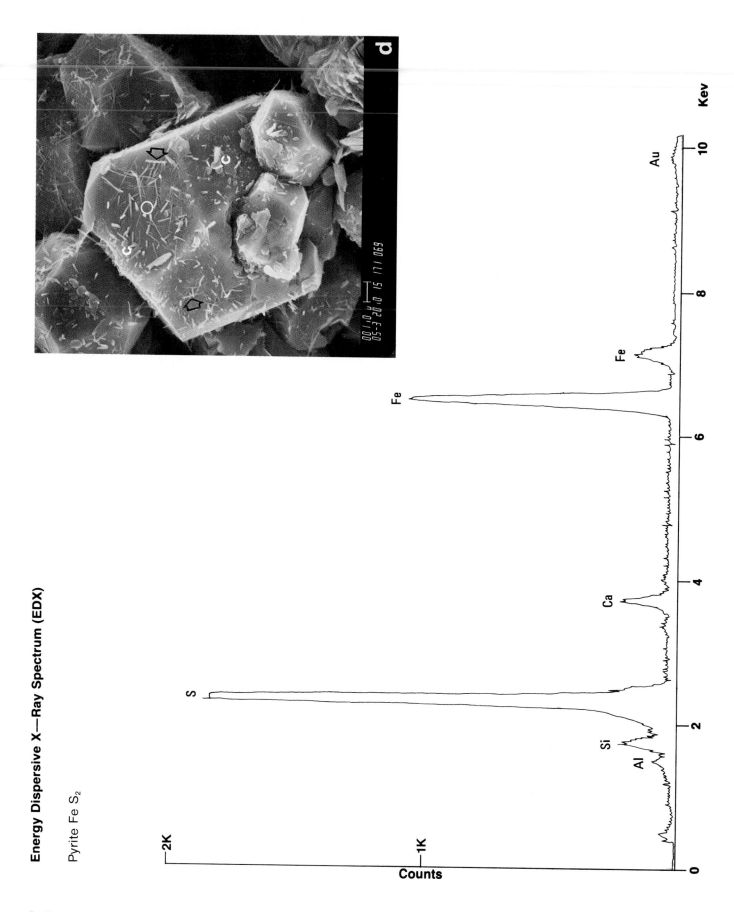

Counts

2K

1K

0 2 4 6 8 10 **Kev**

S

Si

Al

Ca

Fe

Fe

Au

d

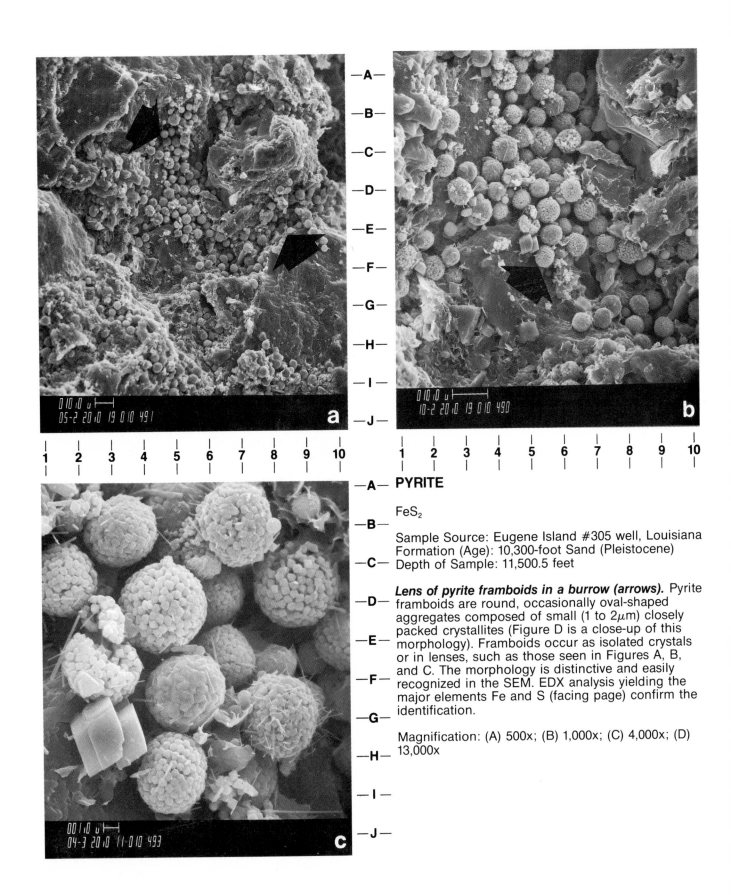

PYRITE

FeS$_2$

Sample Source: Eugene Island #305 well, Louisiana
Formation (Age): 10,300-foot Sand (Pleistocene)
Depth of Sample: 11,500.5 feet

Lens of pyrite framboids in a burrow (arrows). Pyrite
framboids are round, occasionally oval-shaped
aggregates composed of small (1 to 2μm) closely
packed crystallites (Figure D is a close-up of this
morphology). Framboids occur as isolated crystals
or in lenses, such as those seen in Figures A, B,
and C. The morphology is distinctive and easily
recognized in the SEM. EDX analysis yielding the
major elements Fe and S (facing page) confirm the
identification.

Magnification: (A) 500x; (B) 1,000x; (C) 4,000x; (D)
13,000x

Energy Dispersive X—Ray Spectrum (EDX)

Pyrite Fe S_2

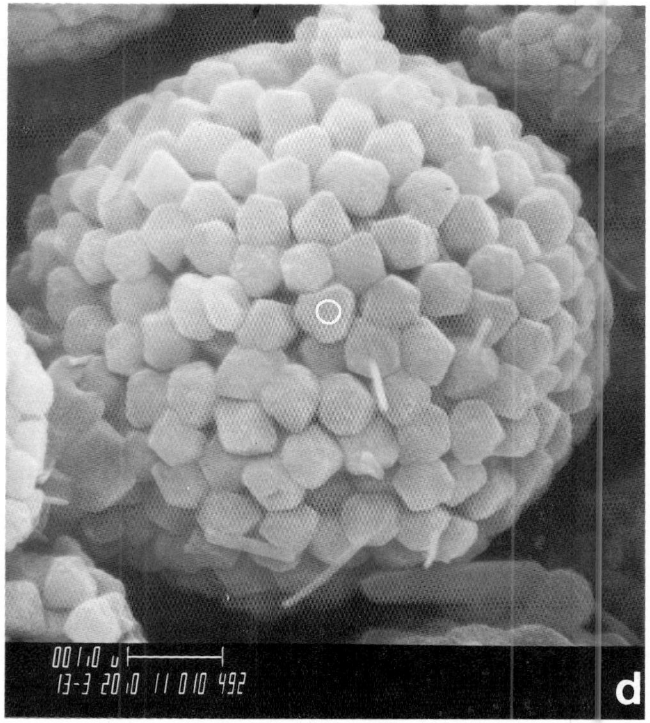
00 1.0 u
13-3 20.0 11 010 492
d

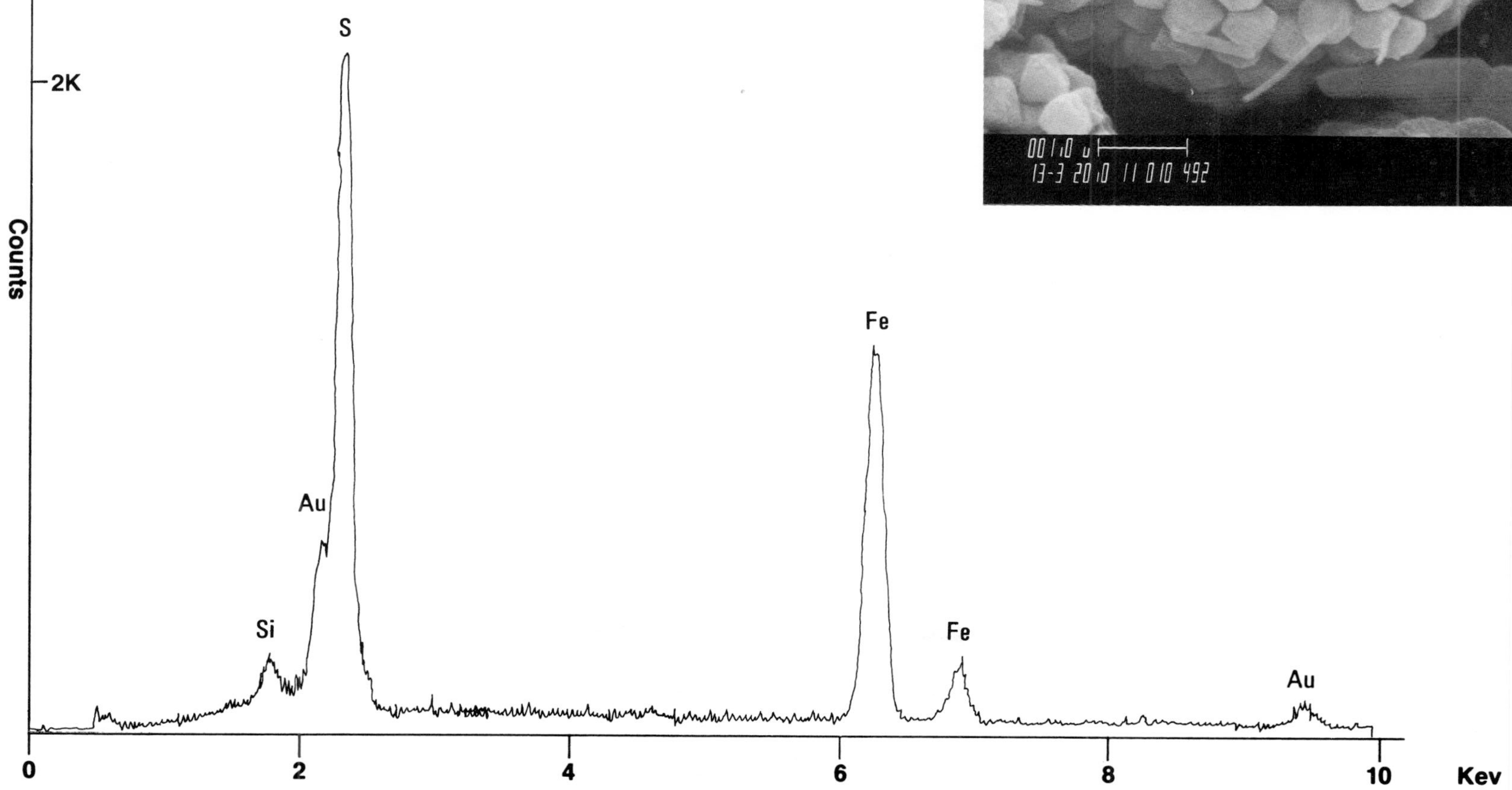
Counts
2K
S
Au
Si
Fe
Fe
Au
0
2
4
6
8
10
Kev

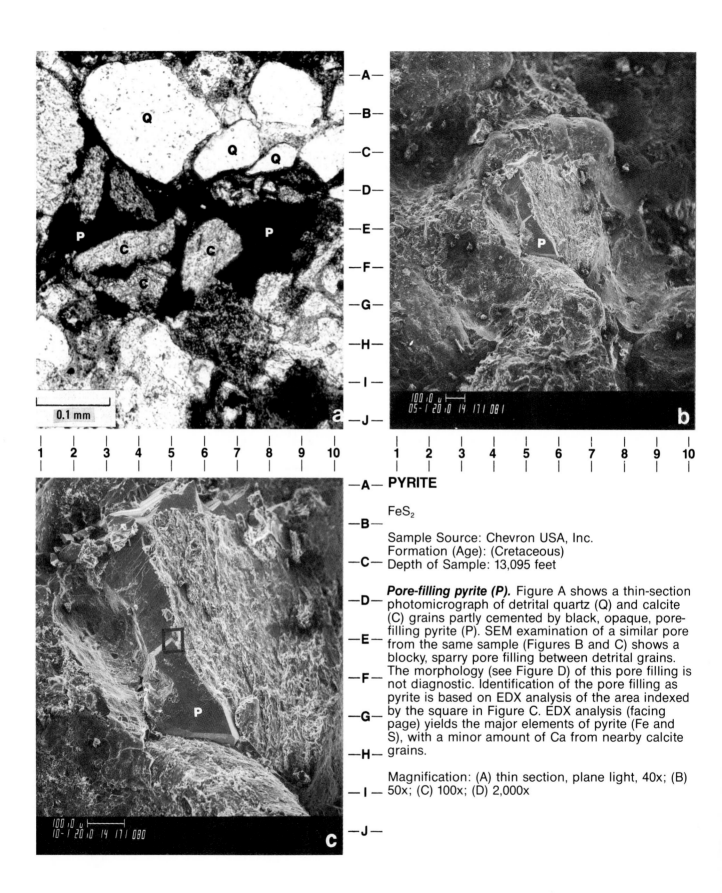

PYRITE

FeS$_2$

Sample Source: Chevron USA, Inc.
Formation (Age): (Cretaceous)
Depth of Sample: 13,095 feet

Pore-filling pyrite (P). Figure A shows a thin-section photomicrograph of detrital quartz (Q) and calcite (C) grains partly cemented by black, opaque, pore-filling pyrite (P). SEM examination of a similar pore from the same sample (Figures B and C) shows a blocky, sparry pore filling between detrital grains. The morphology (see Figure D) of this pore filling is not diagnostic. Identification of the pore filling as pyrite is based on EDX analysis of the area indexed by the square in Figure C. EDX analysis (facing page) yields the major elements of pyrite (Fe and S), with a minor amount of Ca from nearby calcite grains.

Magnification: (A) thin section, plane light, 40x; (B) 50x; (C) 100x; (D) 2,000x

Energy Dispersive X—Ray Spectrum (EDX)

Pyrite Fe S$_2$

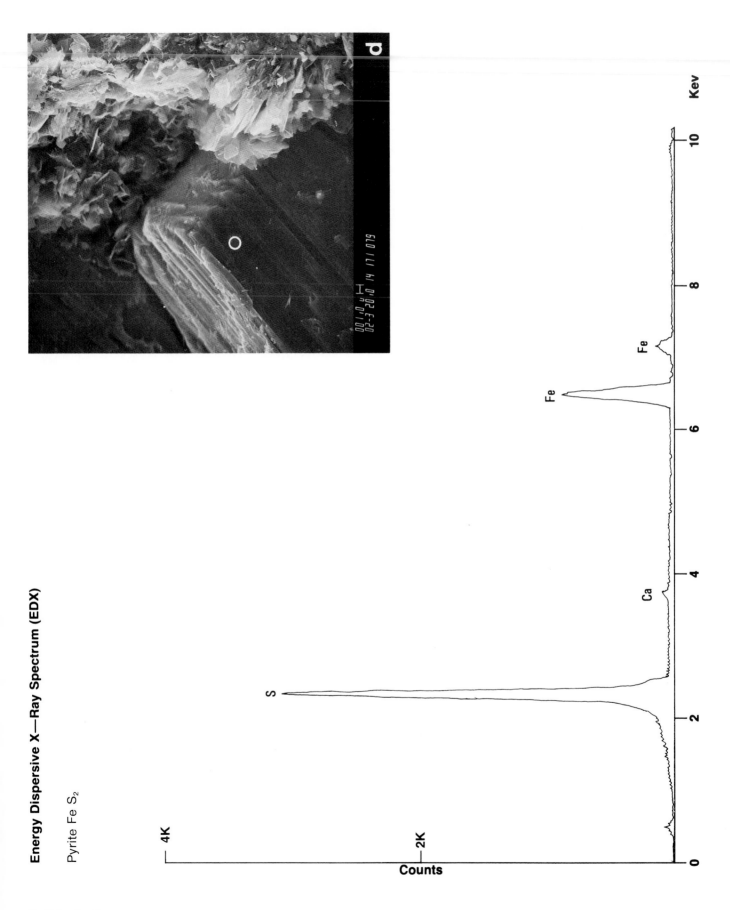

Sulfides-Pyrite

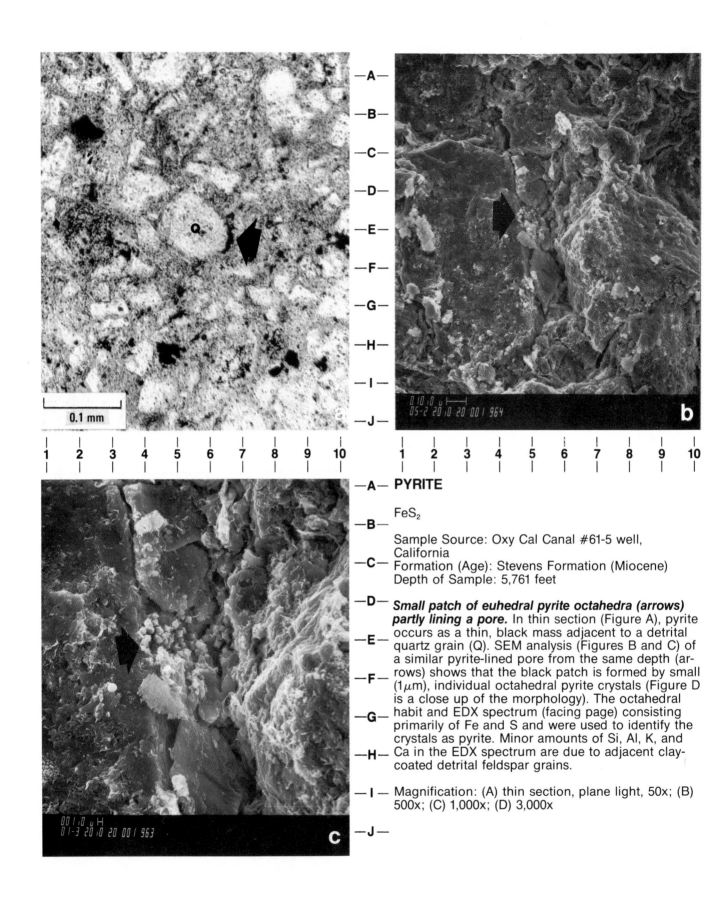

PYRITE

FeS$_2$

Sample Source: Oxy Cal Canal #61-5 well, California
Formation (Age): Stevens Formation (Miocene)
Depth of Sample: 5,761 feet

Small patch of euhedral pyrite octahedra (arrows) partly lining a pore. In thin section (Figure A), pyrite occurs as a thin, black mass adjacent to a detrital quartz grain (Q). SEM analysis (Figures B and C) of a similar pyrite-lined pore from the same depth (arrows) shows that the black patch is formed by small (1 μm), individual octahedral pyrite crystals (Figure D is a close up of the morphology). The octahedral habit and EDX spectrum (facing page) consisting primarily of Fe and S and were used to identify the crystals as pyrite. Minor amounts of Si, Al, K, and Ca in the EDX spectrum are due to adjacent clay-coated detrital feldspar grains.

Magnification: (A) thin section, plane light, 50x; (B) 500x; (C) 1,000x; (D) 3,000x

Energy Dispersive X—Ray Spectrum (EDX)

Pyrite Fe S$_2$

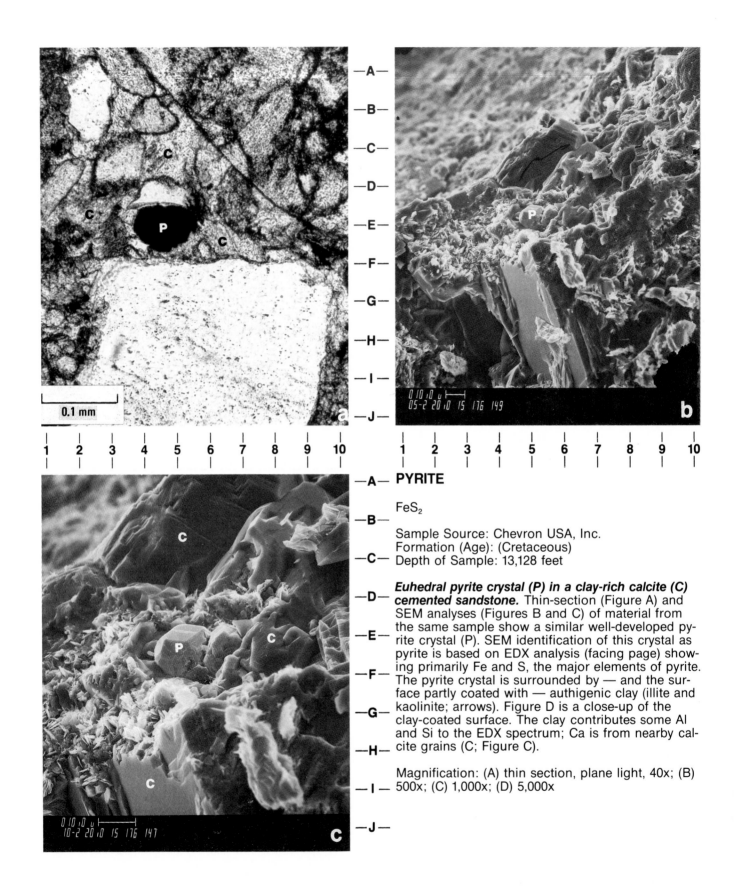

PYRITE

FeS$_2$

Sample Source: Chevron USA, Inc.
Formation (Age): (Cretaceous)
Depth of Sample: 13,128 feet

Euhedral pyrite crystal (P) in a clay-rich calcite (C) cemented sandstone. Thin-section (Figure A) and SEM analyses (Figures B and C) of material from the same sample show a similar well-developed pyrite crystal (P). SEM identification of this crystal as pyrite is based on EDX analysis (facing page) showing primarily Fe and S, the major elements of pyrite. The pyrite crystal is surrounded by — and the surface partly coated with — authigenic clay (illite and kaolinite; arrows). Figure D is a close-up of the clay-coated surface. The clay contributes some Al and Si to the EDX spectrum; Ca is from nearby calcite grains (C; Figure C).

Magnification: (A) thin section, plane light, 40x; (B) 500x; (C) 1,000x; (D) 5,000x

Energy Dispersive X—Ray Spectrum (EDX)

Pyrite Fe S₂

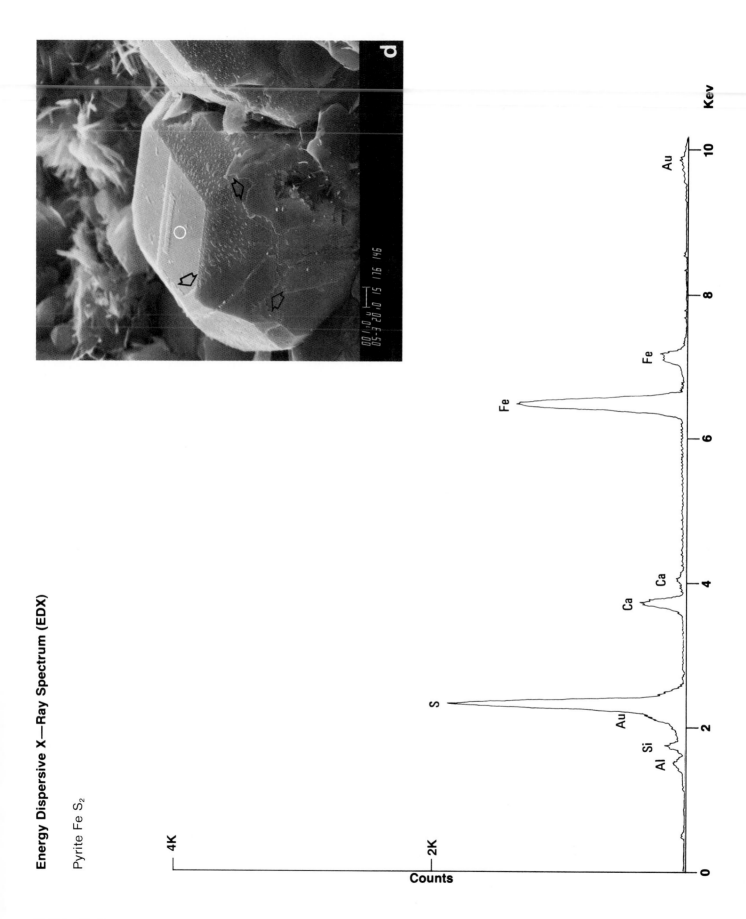

Sulfides-Pyrite

199

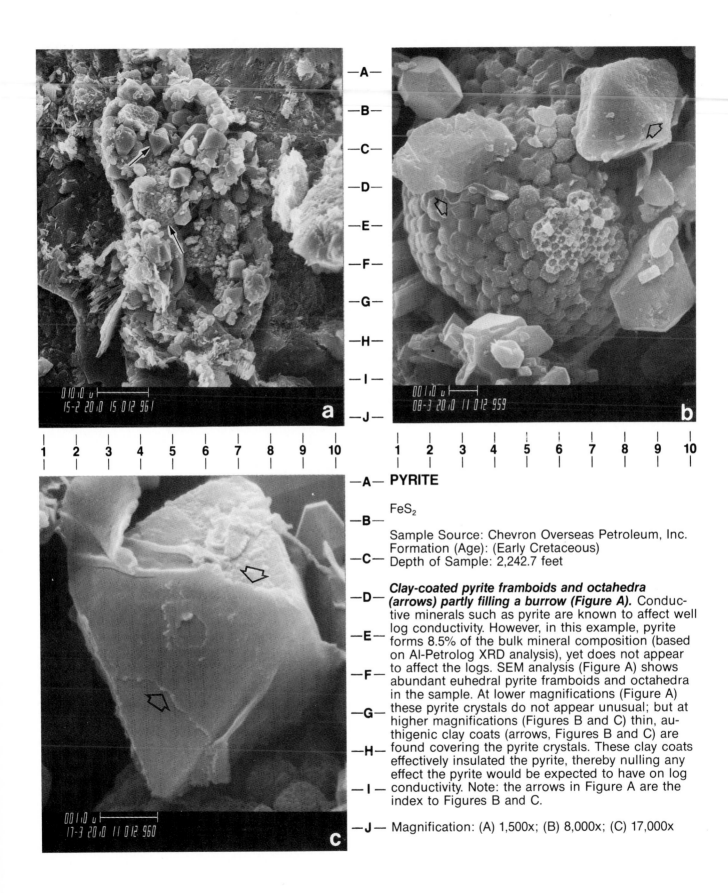

-A- **PYRITE**

FeS$_2$

Sample Source: Chevron Overseas Petroleum, Inc.
Formation (Age): (Early Cretaceous)
Depth of Sample: 2,242.7 feet

Clay-coated pyrite framboids and octahedra (arrows) partly filling a burrow (Figure A). Conductive minerals such as pyrite are known to affect well log conductivity. However, in this example, pyrite forms 8.5% of the bulk mineral composition (based on Al-Petrolog XRD analysis), yet does not appear to affect the logs. SEM analysis (Figure A) shows abundant euhedral pyrite framboids and octahedra in the sample. At lower magnifications (Figure A) these pyrite crystals do not appear unusual; but at higher magnifications (Figures B and C) thin, authigenic clay coats (arrows, Figures B and C) are found covering the pyrite crystals. These clay coats effectively insulated the pyrite, thereby nulling any effect the pyrite would be expected to have on log conductivity. Note: the arrows in Figure A are the index to Figures B and C.

Magnification: (A) 1,500x; (B) 8,000x; (C) 17,000x

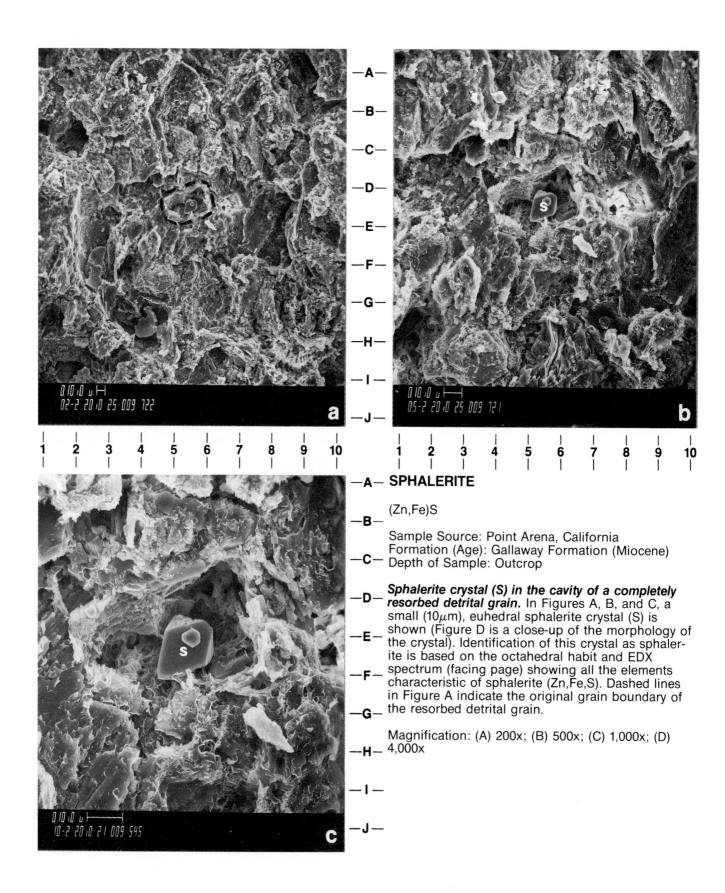

SPHALERITE

(Zn,Fe)S

Sample Source: Point Arena, California
Formation (Age): Gallaway Formation (Miocene)
Depth of Sample: Outcrop

Sphalerite crystal (S) in the cavity of a completely resorbed detrital grain. In Figures A, B, and C, a small (10μm), euhedral sphalerite crystal (S) is shown (Figure D is a close-up of the morphology of the crystal). Identification of this crystal as sphalerite is based on the octahedral habit and EDX spectrum (facing page) showing all the elements characteristic of sphalerite (Zn,Fe,S). Dashed lines in Figure A indicate the original grain boundary of the resorbed detrital grain.

Magnification: (A) 200x; (B) 500x; (C) 1,000x; (D) 4,000x

Energy Dispersive X—Ray Spectrum (EDX)

Sphalerite (Zn, Fe) S

GYPSUM (ARTIFICIAL)

$CaSO_4 \cdot 2 H_2O$

Sample Source: Panna Maria, Texas
Formation (Age): Jackson Formation (Eocene)
Depth of Sample: Outcrop

Gypsum rosettes (G) formed of lath-like gypsum crystals. In Figures A, B, and C, short prismatic laths (Figure A, coordinates E5; Figure B, D5; and Figure C, C8) and long prismatic laths (Figure A, E4; Figure B, G7; and Figure C, C3) are observed arranged into a rosette (long laths range from 5 to 30μm long). Identification of these crystals as gypsum is based on correlating the morphology with the EDX analysis (facing page) which indicates the major elements of gypsum (nearly equal amounts of Ca and S). The EDX spectrum is identical to anhydrite, so identification is based on crystal habit and supplemented by XRD analysis. These rosettes were artificially formed during experiments by C. McAllister. Sample courtesy of P. Kimbrell.

Magnification: (A) 300x; (B) 700x; (C) 2,000x; (D) 10,000x

Energy Dispersive X—Ray Spectrum (EDX)

Gypsum Ca So$_4$ • 2 H$_2$O

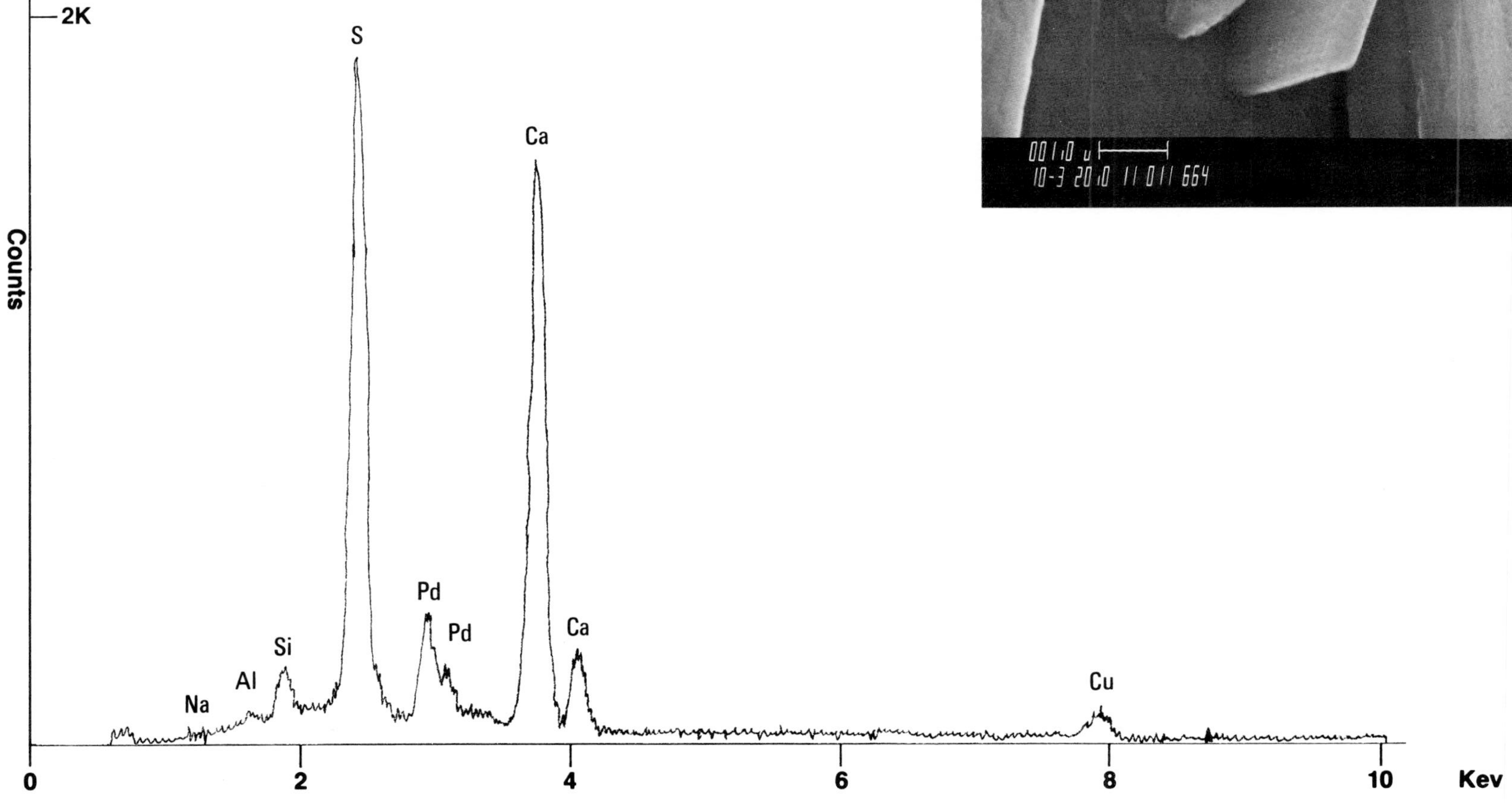

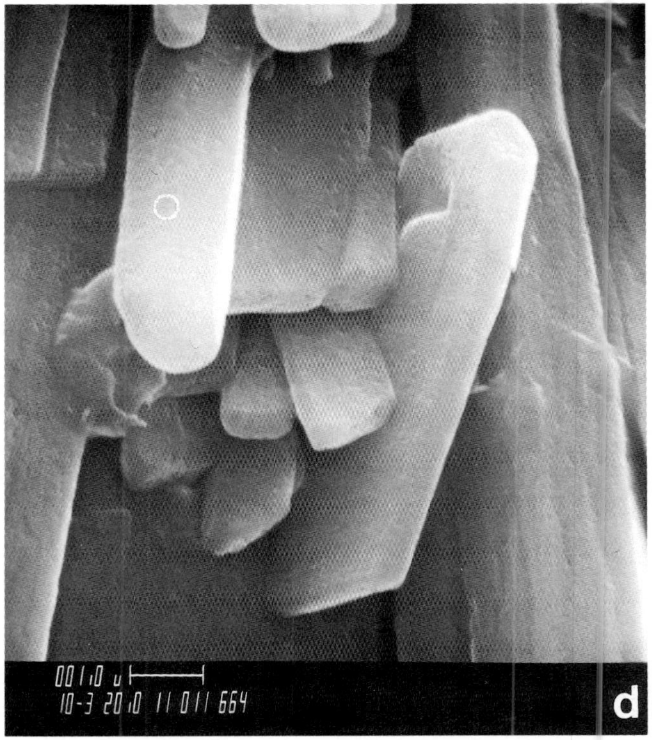

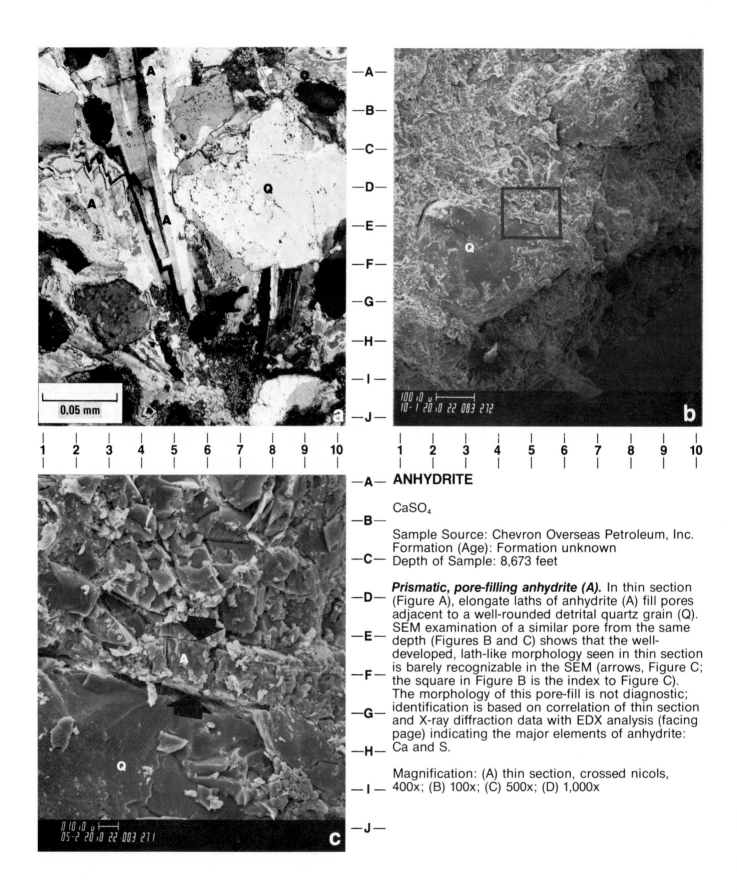

ANHYDRITE

CaSO$_4$

Sample Source: Chevron Overseas Petroleum, Inc.
Formation (Age): Formation unknown
Depth of Sample: 8,673 feet

Prismatic, pore-filling anhydrite (A). In thin section (Figure A), elongate laths of anhydrite (A) fill pores adjacent to a well-rounded detrital quartz grain (Q). SEM examination of a similar pore from the same depth (Figures B and C) shows that the well-developed, lath-like morphology seen in thin section is barely recognizable in the SEM (arrows, Figure C; the square in Figure B is the index to Figure C). The morphology of this pore-fill is not diagnostic; identification is based on correlation of thin section and X-ray diffraction data with EDX analysis (facing page) indicating the major elements of anhydrite: Ca and S.

Magnification: (A) thin section, crossed nicols, 400x; (B) 100x; (C) 500x; (D) 1,000x

Energy Dispersive X—Ray Spectrum (EDX)

Anhydrite Ca SO₄

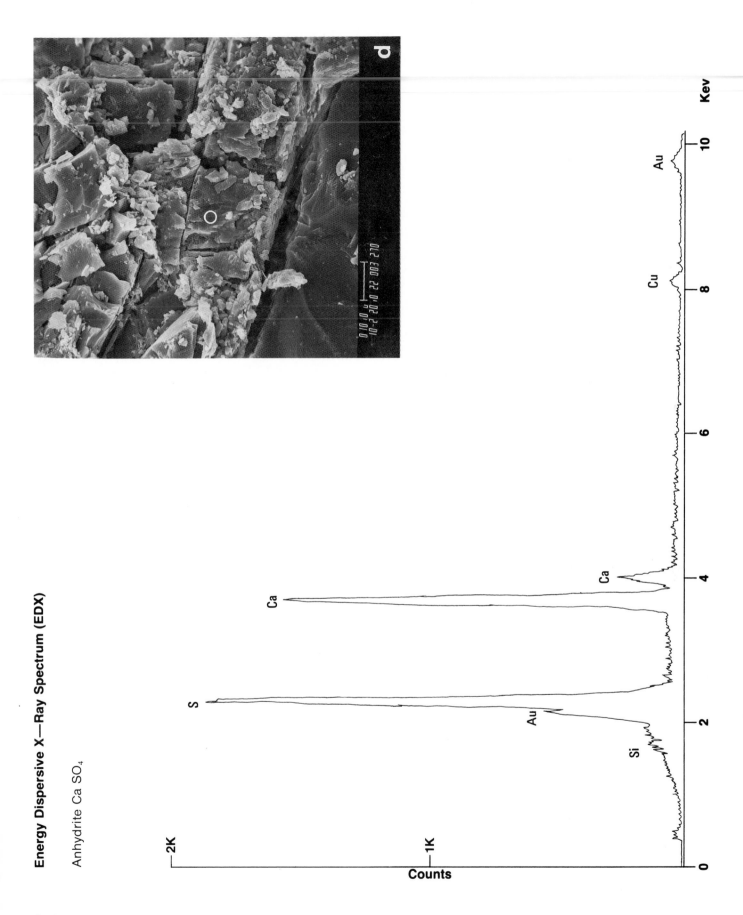

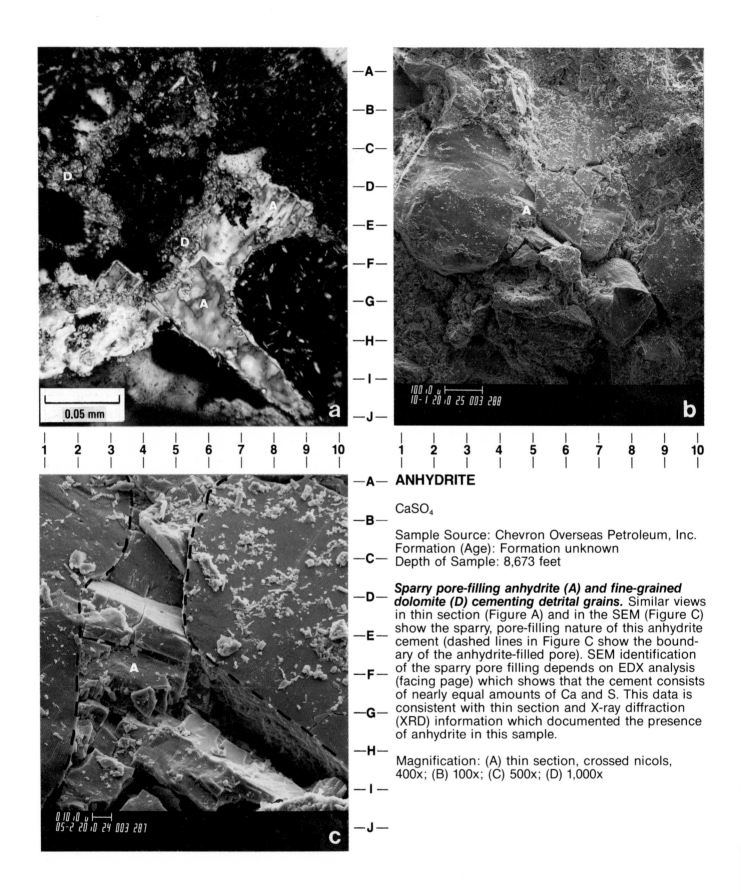

ruler scale: 1 2 3 4 5 6 7 8 9 10

ANHYDRITE

CaSO$_4$

Sample Source: Chevron Overseas Petroleum, Inc.
Formation (Age): Formation unknown
Depth of Sample: 8,673 feet

Sparry pore-filling anhydrite (A) and fine-grained dolomite (D) cementing detrital grains. Similar views in thin section (Figure A) and in the SEM (Figure C) show the sparry, pore-filling nature of this anhydrite cement (dashed lines in Figure C show the boundary of the anhydrite-filled pore). SEM identification of the sparry pore filling depends on EDX analysis (facing page) which shows that the cement consists of nearly equal amounts of Ca and S. This data is consistent with thin section and X-ray diffraction (XRD) information which documented the presence of anhydrite in this sample.

Magnification: (A) thin section, crossed nicols, 400x; (B) 100x; (C) 500x; (D) 1,000x

Energy Dispersive X—Ray Spectrum (EDX)

Anhydrite Ca SO₄

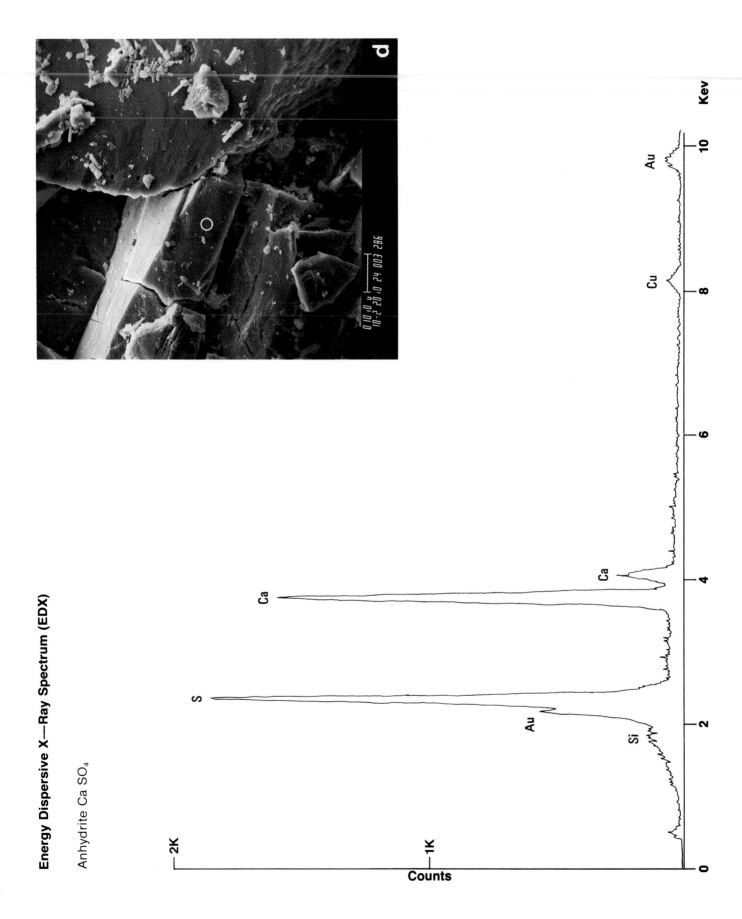

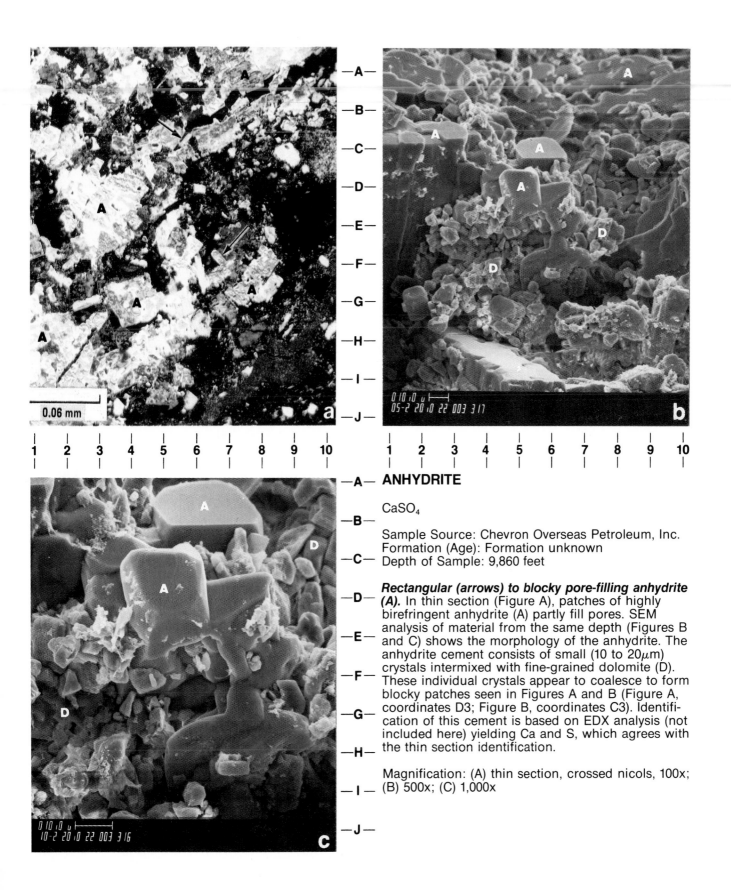

ANHYDRITE

CaSO$_4$

Sample Source: Chevron Overseas Petroleum, Inc.
Formation (Age): Formation unknown
Depth of Sample: 9,860 feet

Rectangular (arrows) to blocky pore-filling anhydrite (A). In thin section (Figure A), patches of highly birefringent anhydrite (A) partly fill pores. SEM analysis of material from the same depth (Figures B and C) shows the morphology of the anhydrite. The anhydrite cement consists of small (10 to 20μm) crystals intermixed with fine-grained dolomite (D). These individual crystals appear to coalesce to form blocky patches seen in Figures A and B (Figure A, coordinates D3; Figure B, coordinates C3). Identification of this cement is based on EDX analysis (not included here) yielding Ca and S, which agrees with the thin section identification.

Magnification: (A) thin section, crossed nicols, 100x; (B) 500x; (C) 1,000x

A— **COPIAPITE**

B— (Fe,Mg)Fe$_4^{+3}$(SO$_4$)$_6$(OH)$_2$ • 20 H$_2$O

Sample Source: Ward's Pyrite Standard, Custer, South Dakota
C— Formation (Age): Formation unknown
Depth of Sample: Outcrop

D— ***Thin, tabular crystals of copiapite (C) encrusting pyrite (P).*** Copiapite forms by the oxidation of pyrite. Figure A shows a pyrite sample (P) coated
E— with a blobby, discontinuous crust of copiapite crystals (C; Figures B and C). Individual crystals of copiapite are small (5 to 10μm), tabular, euhedral,
F— and arranged on-edge to the pyrite surface (Figure D is a close-up of the morphology). The EDX spectrum (facing page) is similar to pyrite, so identifica-
G— tion must be supplemented by comparison of the morphology and mineral associations. Sample courtesy of S. McDonald; identification by A. Carpenter.
H—

Magnification: (A) 10x; (B) 100x; (C) 500x; (D) 1,000x

Energy Dispersive X—Ray Spectrum (EDX)

Copiapite (Fe, Mg) Fe_4^{+3} $(SO_4)_6$ $(OH)_2 \cdot 20\ H_2O$

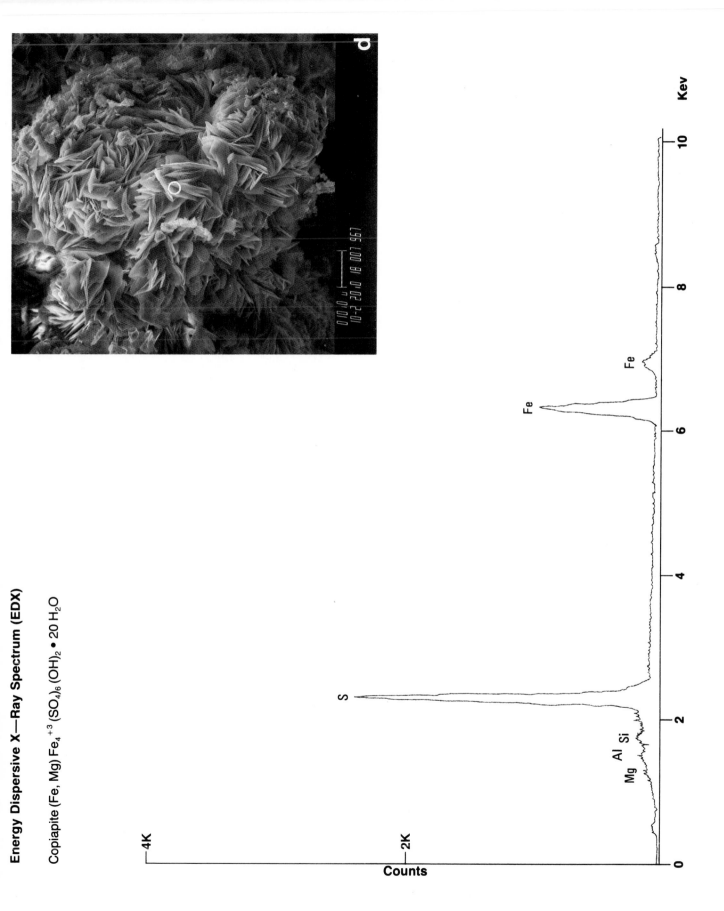

Counts

Kev

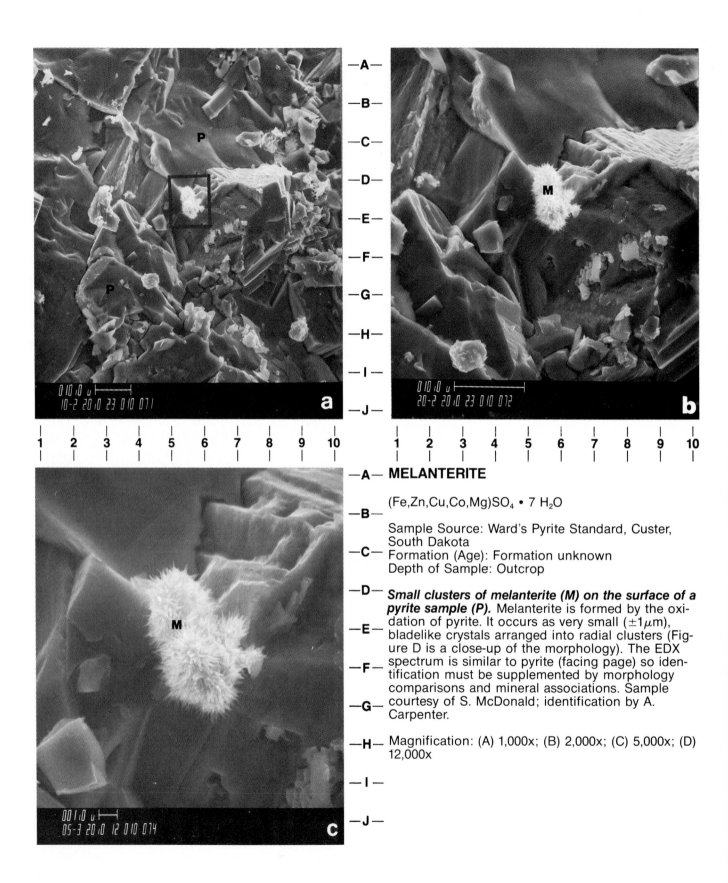

MELANTERITE

$(Fe,Zn,Cu,Co,Mg)SO_4 \cdot 7\, H_2O$

Sample Source: Ward's Pyrite Standard, Custer, South Dakota
Formation (Age): Formation unknown
Depth of Sample: Outcrop

Small clusters of melanterite (M) on the surface of a pyrite sample (P). Melanterite is formed by the oxidation of pyrite. It occurs as very small ($\pm 1\mu$m), bladelike crystals arranged into radial clusters (Figure D is a close-up of the morphology). The EDX spectrum is similar to pyrite (facing page) so identification must be supplemented by morphology comparisons and mineral associations. Sample courtesy of S. McDonald; identification by A. Carpenter.

Magnification: (A) 1,000x; (B) 2,000x; (C) 5,000x; (D) 12,000x

Energy Dispersive X—Ray Spectrum (EDX)

Melanterite (Fe, Zn, Cu, Co, Mg) $SO_4 \cdot 7\ H_2O$

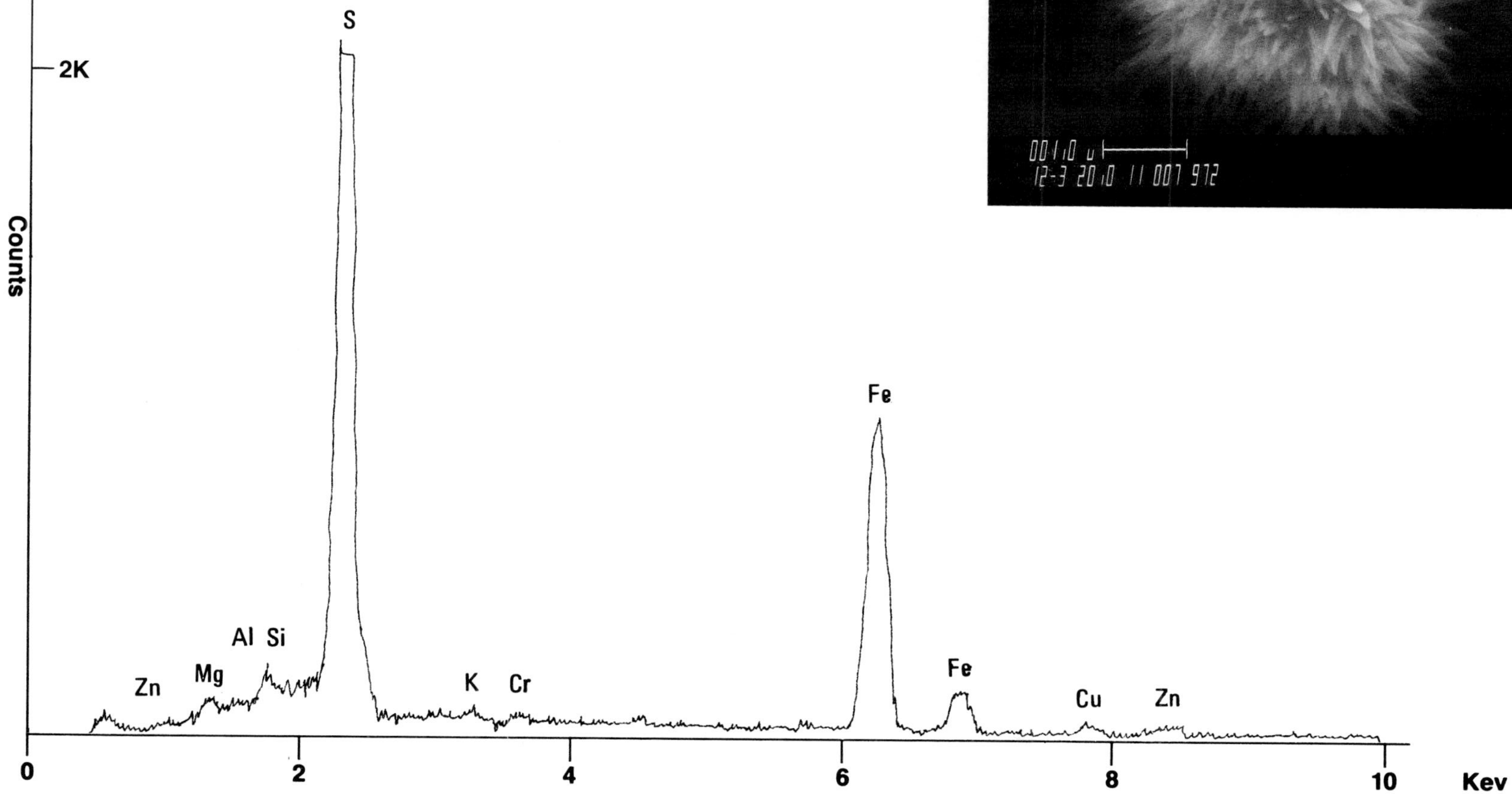

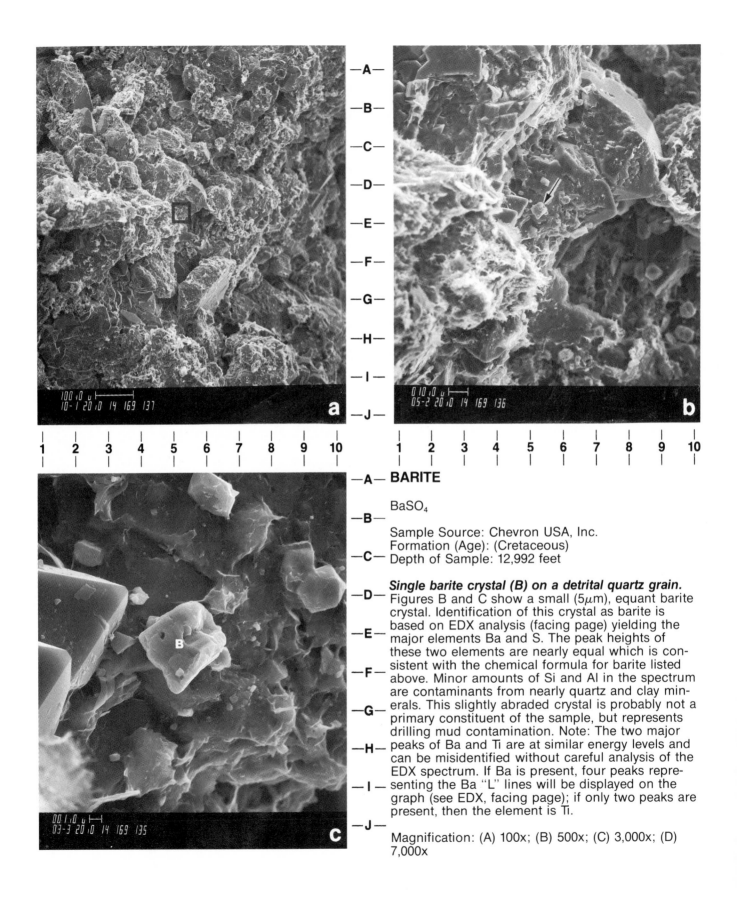

BARITE

BaSO$_4$

Sample Source: Chevron USA, Inc.
Formation (Age): (Cretaceous)
Depth of Sample: 12,992 feet

Single barite crystal (B) on a detrital quartz grain.
Figures B and C show a small (5μm), equant barite crystal. Identification of this crystal as barite is based on EDX analysis (facing page) yielding the major elements Ba and S. The peak heights of these two elements are nearly equal which is consistent with the chemical formula for barite listed above. Minor amounts of Si and Al in the spectrum are contaminants from nearly quartz and clay minerals. This slightly abraded crystal is probably not a primary constituent of the sample, but represents drilling mud contamination. Note: The two major peaks of Ba and Ti are at similar energy levels and can be misidentified without careful analysis of the EDX spectrum. If Ba is present, four peaks representing the Ba "L" lines will be displayed on the graph (see EDX, facing page); if only two peaks are present, then the element is Ti.

Magnification: (A) 100x; (B) 500x; (C) 3,000x; (D) 7,000x

Energy Dispersive X—Ray Spectrum (EDX)

Barite Ba SO$_4$

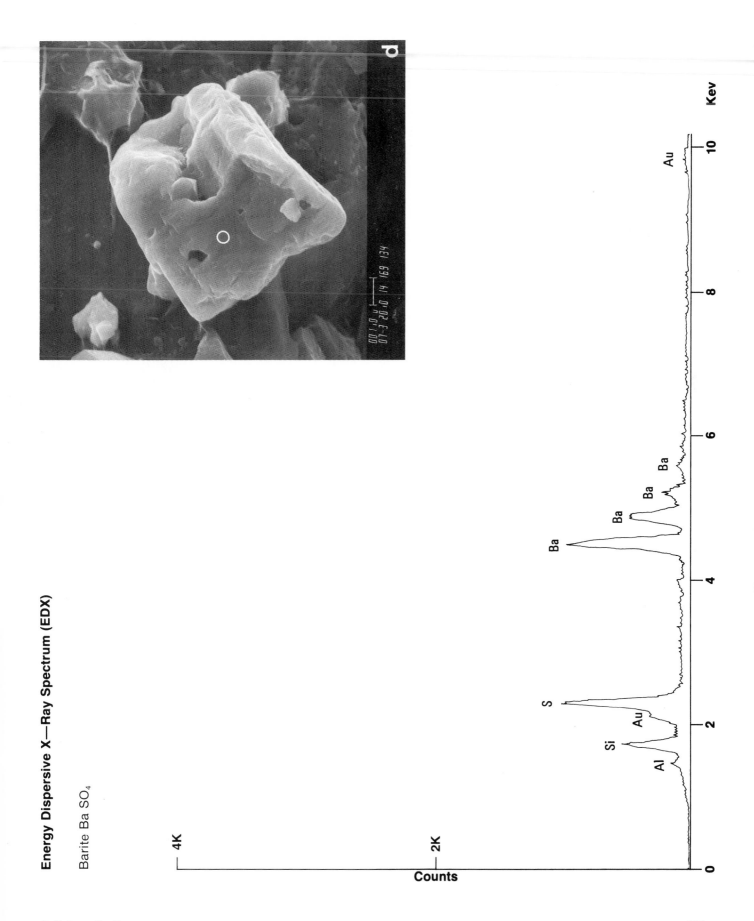

A

B

C

D

E

F

G

H

I

J

1 2 3 4 5 6 7 8 9 10

1 2 3 4 5 6 7 8 9 10

HEMATITE AND GOETHITE

Fe_2O_3 and $FeO(OH)$

Sample Source: Southeast Missouri
Formation (Age): Taum Sauk Limestone (Late Cambrian)
Depth of Sample: Outcrop

Hematite (H) and goethite (G) inclusions within a partly dedolomitized dolomite rhomb (D). Figure A shows a large (1mm), euhedral, dolomite rhomb, partly altered to calcite. The calcite was dissolved using HCl, revealing small inclusions (arrow, Figure B) filled with iron oxides (G and H, Figure C); the square in Figure A indexes Figure B. Two crystal habits of iron oxide are seen in the dolomite: a small (less than 1μm), round or disc-shaped hematite (?) (see H, Figure C, coordinates E7); and a radiating, rod-shaped goethite (?) (see G, Figure C, coordinates F5). These minute iron oxide inclusions are responsible for the reddish-brown coloration of the dolomite crystals. For additional examples see Frank (1981). Sample and identification courtesy of J. Frank and A. Carpenter.

Magnification: (A) 60x; (B) 3,000x; (C) 15,000x

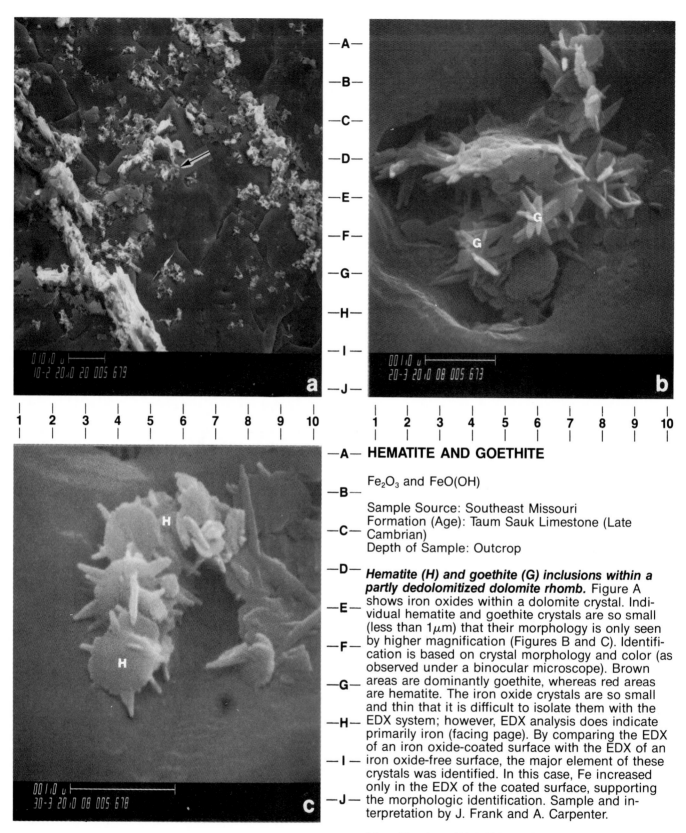

HEMATITE AND GOETHITE

Fe$_2$O$_3$ and FeO(OH)

Sample Source: Southeast Missouri
Formation (Age): Taum Sauk Limestone (Late Cambrian)
Depth of Sample: Outcrop

Hematite (H) and goethite (G) inclusions within a partly dedolomitized dolomite rhomb. Figure A shows iron oxides within a dolomite crystal. Individual hematite and goethite crystals are so small (less than 1μm) that their morphology is only seen by higher magnification (Figures B and C). Identification is based on crystal morphology and color (as observed under a binocular microscope). Brown areas are dominantly goethite, whereas red areas are hematite. The iron oxide crystals are so small and thin that it is difficult to isolate them with the EDX system; however, EDX analysis does indicate primarily iron (facing page). By comparing the EDX of an iron oxide-coated surface with the EDX of an iron oxide-free surface, the major element of these crystals was identified. In this case, Fe increased only in the EDX of the coated surface, supporting the morphologic identification. Sample and interpretation by J. Frank and A. Carpenter.

Magnification: (A) 1,000x; (B) 20,000x; (C) 30,000x; (D) 20,000x

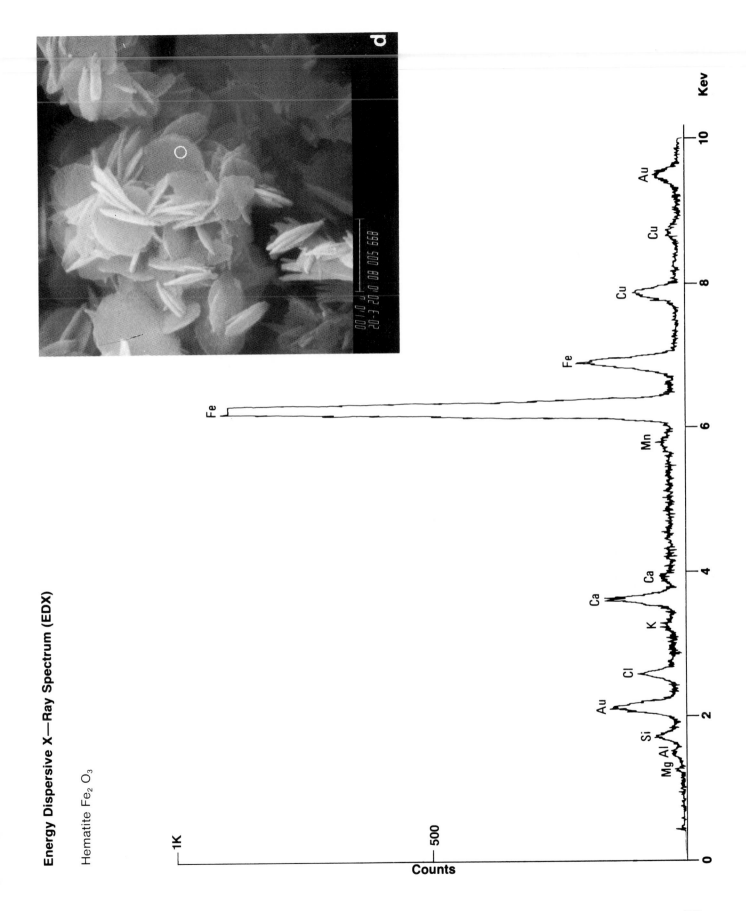

Energy Dispersive X—Ray Spectrum (EDX)

Hematite Fe_2O_3

Oxides—Hematite

221

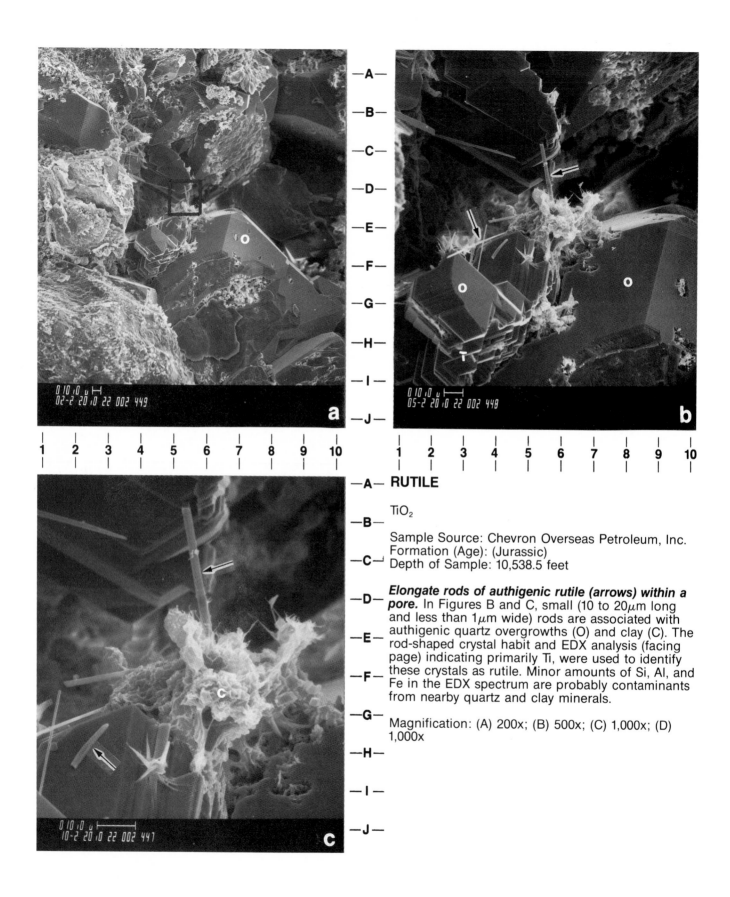

RUTILE

TiO$_2$

Sample Source: Chevron Overseas Petroleum, Inc.
Formation (Age): (Jurassic)
Depth of Sample: 10,538.5 feet

Elongate rods of authigenic rutile (arrows) within a pore. In Figures B and C, small (10 to 20μm long and less than 1μm wide) rods are associated with authigenic quartz overgrowths (O) and clay (C). The rod-shaped crystal habit and EDX analysis (facing page) indicating primarily Ti, were used to identify these crystals as rutile. Minor amounts of Si, Al, and Fe in the EDX spectrum are probably contaminants from nearby quartz and clay minerals.

Magnification: (A) 200x; (B) 500x; (C) 1,000x; (D) 1,000x

Energy Dispersive X—Ray Spectrum (EDX)

Rutile Ti O_2

Energy Dispersive X—Ray Spectrum (EDX)

Magnetite $Fe^{+2} Fe^{+3}_2 O_4$

Energy Dispersive X—Ray Spectrum (EDX)

Ilmenite Fe Ti O₃

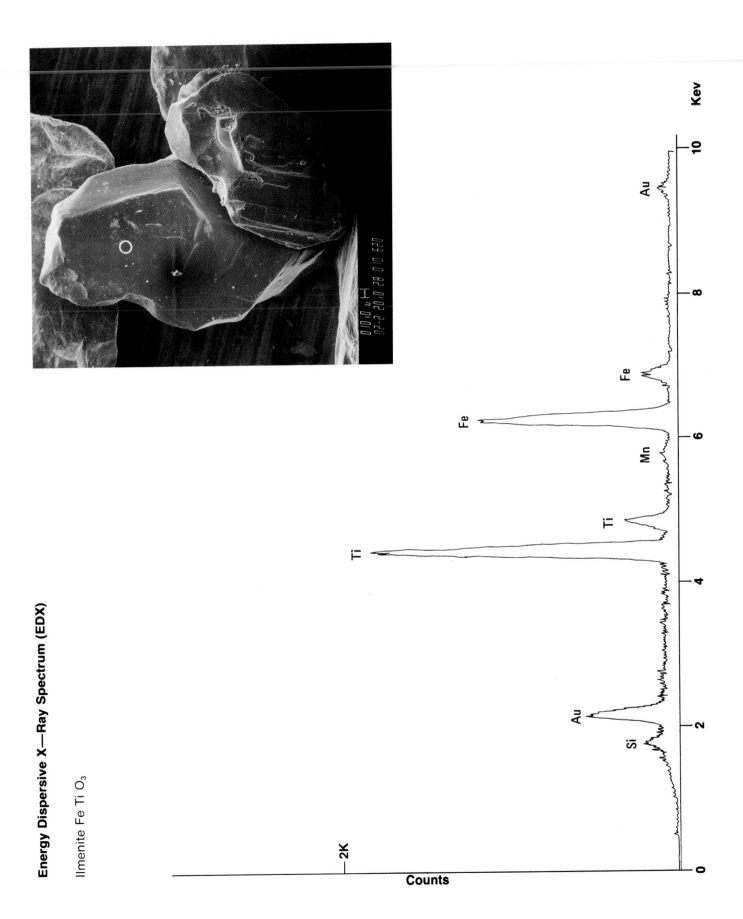

Counts

2K

Si
Au
Ti
Ti
Mn
Fe
Fe
Au

0   2   4   6   8   10   **Kev**

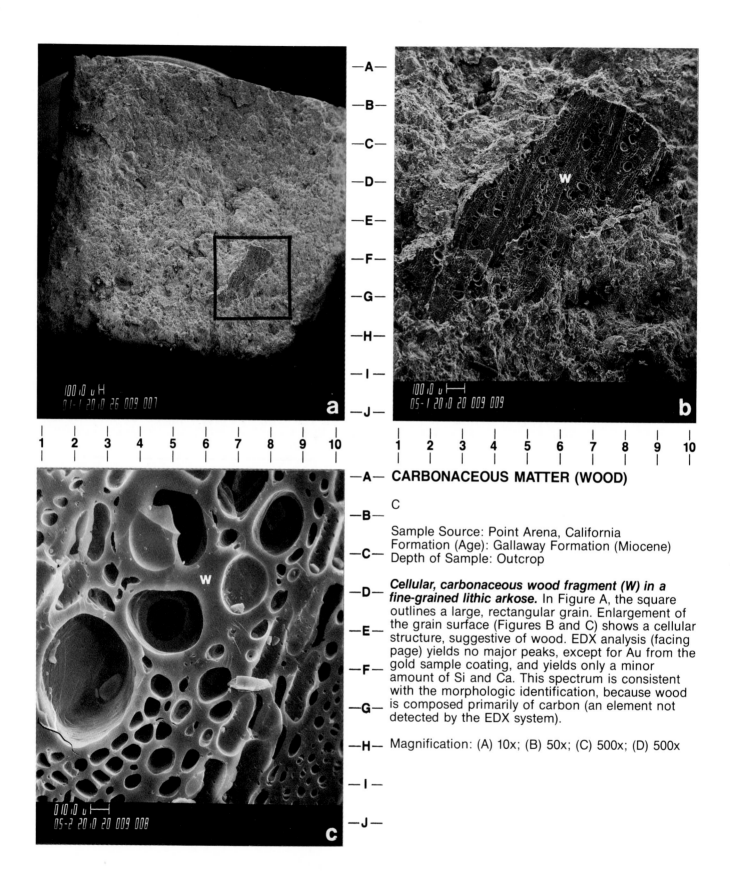

CARBONACEOUS MATTER (WOOD)

C

Sample Source: Point Arena, California
Formation (Age): Gallaway Formation (Miocene)
Depth of Sample: Outcrop

Cellular, carbonaceous wood fragment (W) in a fine-grained lithic arkose. In Figure A, the square outlines a large, rectangular grain. Enlargement of the grain surface (Figures B and C) shows a cellular structure, suggestive of wood. EDX analysis (facing page) yields no major peaks, except for Au from the gold sample coating, and yields only a minor amount of Si and Ca. This spectrum is consistent with the morphologic identification, because wood is composed primarily of carbon (an element not detected by the EDX system).

Magnification: (A) 10x; (B) 50x; (C) 500x; (D) 500x

Energy Dispersive X—Ray Spectrum (EDX)

Carbonaceous Fragment (Wood) C

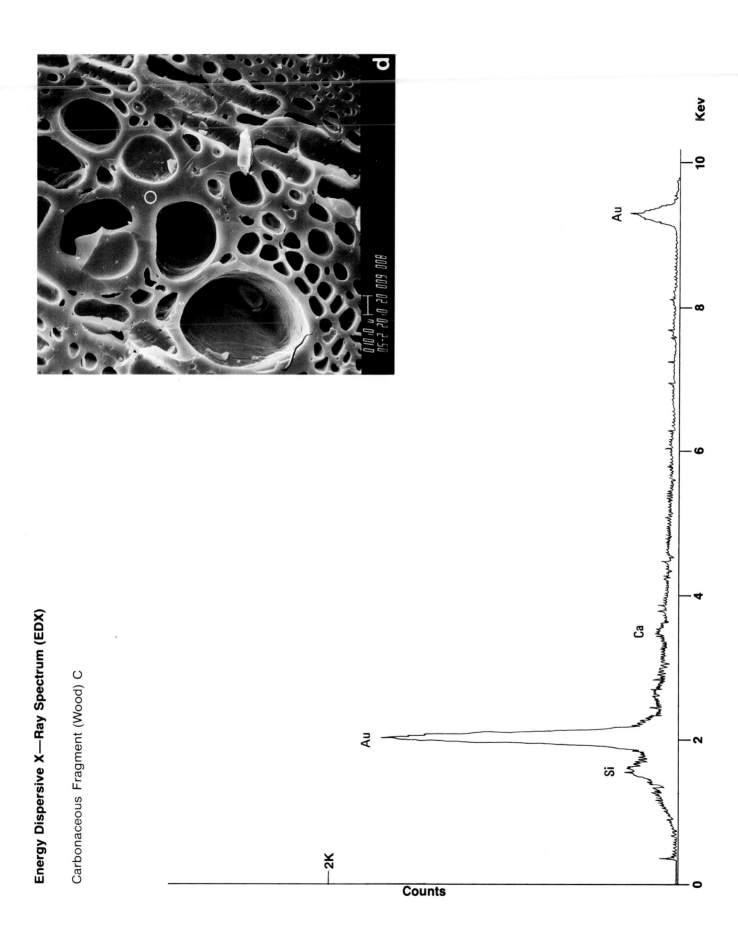

Energy Table for Characteristic X-Ray Transitions

Energies are given in keV. Columns are grouped into K-SERIES, L-SERIES (L_I, L_{II}, L_{III} sub-series) and M-SERIES.

Z Element	$K(ab)$	K_{β_3}	K_{β_1}	K_{β_2}	K_{α_1}	K_{α_2}	$L_I(ab)$	L_{γ_3}	L_{γ_2}	L_{β_3}	L_{β_4}	$L_{II}(ab)$	L_{γ_1}	L_{β_1}	L_η	$L_{III}(ab)$	L_{β_2}	L_{α_1}	L_{α_2}	L_L	$M_V(ab)$	$M_{IV}(ab)$	M_β	M_{α_1}	M_{α_2}
1 H	0.0136																								
2 He	0.025																								
3 Li	0.055				0.052	0.052																			
4 Be	0.112				0.110	0.110																			
5 B	0.192				0.185	0.185																			
6 C	0.283				0.277	0.277																			
7 N	0.399				0.392	0.392																			
8 O	0.531				0.525	0.525																			
9 F	0.687				0.677	0.677																			
10 Ne	0.867				0.848	0.848																			
11 Na	1.072		1.067		1.041	1.041																			
12 Mg	1.305		1.295		1.253	1.253																			
13 Al	1.559		1.553		1.486	1.486																			
14 Si	1.838		1.836		1.740	1.739																			
15 P	2.142		2.136		2.013	2.012																			
16 S	2.472		2.464		2.307	2.306																			
17 Cl	2.822				2.622	2.620																			
18 Ar	3.202		3.190		2.957	2.955																			
19 K	3.607		3.589		3.313	3.310									0.262					0.260					
20 Ca	4.038		4.012		3.691	3.687	0.400					0.350		0.345	0.306	0.346		0.341		0.303					
21 Sc	4.496		4.460		4.090	4.085	0.463					0.407		0.400	0.353	0.403		0.395		0.348					
22 Ti	4.965		4.931		4.510	4.504	0.530					0.460		0.458	0.401	0.454		0.452		0.395					
23 V	5.465		5.426		4.951	4.944	0.604			0.585		0.520		0.519	0.453	0.513		0.511		0.446					
24 Cr	5.989		5.946		5.414	5.405	0.682			0.654		0.583		0.583	0.510	0.574		0.573		0.500					
25 Mn	6.540		6.489		5.898	5.887	0.754			0.721		0.652		0.649	0.567	0.641		0.637		0.556					
26 Fe	7.112		7.057		6.403	6.390	0.842			0.792		0.721		0.718	0.628	0.709		0.705		0.615					
27 Co	7.709		7.648		6.929	6.914	0.929			0.866		0.794		0.791	0.694	0.779		0.776		0.678					
28 Ni	8.333		8.263		7.477	7.460	1.012			0.941		0.872		0.869	0.762	0.855		0.851		0.743					
29 Cu	8.979		8.904		8.046	8.026	1.100			1.023		0.952		0.950	0.832	0.932		0.930		0.811					
30 Zn	9.659		9.570	9.656	8.637	8.614	1.196			1.107		1.044		1.034	0.906	1.021		1.012		0.884					
31 Ga	10.368	10.259	10.263	10.365	9.250	9.223	1.300				1.197	1.134		1.125	0.984	1.117		1.098		0.957					
32 Ge	11.104	10.976	10.980	11.099	9.885	9.854	1.420				1.286	1.249		1.218	1.068	1.218		1.188		1.036					
33 As	11.868	11.718	11.724	11.862	10.542	10.506	1.530				1.388	1.360		1.317	1.155	1.325		1.282		1.120					
34 Se	12.658	12.437	12.494	12.650	11.220	11.179	1.653			1.706	1.490	1.477		1.419	1.244	1.436		1.379		1.204					
35 Br	13.474	13.282	13.289	13.467	11.922	11.876	1.794				1.596	1.596		1.526	1.339	1.550		1.480		1.293					
36 Kr	14.322	14.102	14.110	14.312	12.648	12.596	1.920				1.697	1.756		1.636		1.675		1.586							
37 Rb	15.201	14.949	14.959	15.183	13.393	13.333	2.067			1.826	1.817	1.866		1.752	1.542	1.806		1.694	1.692	1.482					
38 Sr	16.105	15.822	15.833	16.082	14.163	14.095	2.216			1.947	1.936	2.007		1.871	1.649	1.940		1.806	1.804	1.582					
39 Y	17.037	16.723	16.735	17.013	14.956	14.880	2.369			2.072	2.060	2.145		1.995	1.761	2.079		1.922	1.920	1.685					
40 Zr	17.998	17.651	17.665	17.967	15.772	15.688	2.547			2.201	2.187	2.307	2.302	2.124	1.876	2.223	2.219	2.042	2.040	1.792					
41 Nb	18.986	18.603	18.619	18.949	16.612	16.518	2.698			2.334	2.319	2.465	2.461	2.257	1.996	2.371	2.367	2.166	2.163	1.902					
42 Mo	20.002	19.587	19.605	19.962	17.476	17.371	2.866			2.473	2.455	2.625	2.623	2.394	2.120	2.520	2.518	2.293	2.289	2.015					
43 Tc	21.054	20.595	20.619	21.002	18.364	18.248	3.054			2.763	2.741	2.795		2.536	2.382	2.677		2.424							
44 Ru	22.118	21.631	21.657	22.070	19.276	19.147	3.236			2.763	2.741	2.966	2.964	2.683	2.382	2.837	2.835	2.558	2.554	2.252					
45 Rh	23.224	22.695	22.724	23.169	20.213	20.070	3.419			2.915	2.890	3.146	3.143	2.834	2.519	3.003	3.001	2.696	2.692	2.376					
46 Pd	24.350	23.787	23.819	24.295	21.174	21.017	3.617			3.072	3.045	3.330	3.328	2.990	2.660	3.173	3.171	2.838	2.833	2.503					
47 Ag	25.514	24.907	24.943	25.452	22.159	21.987	3.806		3.749	3.234	3.203	3.524	3.519	3.150	2.806	3.351	3.347	2.984	2.978	2.633					
48 Cd	26.711	26.057	26.096	26.639	23.170	22.980	4.019			3.401	3.367	3.727	3.716	3.316	2.956	3.537	3.528	3.133	3.126	2.767					
49 In	27.940	27.233	27.276	27.856	24.206	23.998	4.237			3.572	3.535	3.938	3.920	3.487	3.112	3.730	3.713	3.286	3.279	2.904					
50 Sn	29.200	28.439	28.486	29.104	25.267	25.040	4.465			3.750	3.708	4.156	4.130	3.662	3.272	3.929	3.904	3.443	3.435	3.044					
51 Sb	30.491	29.674	29.726	30.388	26.355	26.106	4.698			3.932	3.886	4.381	4.347	3.843	3.436	4.132	4.100	3.604	3.595	3.188					
52 Te	31.813	30.939	30.995	31.698	27.468	27.202	4.939			4.120	4.069	4.612	4.570	4.029	3.605	4.341	4.301	3.769	3.758	3.335					
53 I	33.169	32.234	32.295	33.036	28.607	28.317	5.188			4.313	4.257	4.852	4.800	4.220	3.780	4.557	4.507	3.937	3.925	3.484					
54 Xe	34.582	33.556	33.624	34.408	29.774	29.461	5.452					5.100				4.781		4.109							
55 Cs	35.959	34.913	34.987	35.815	30.968	30.623	5.720		5.552	4.716	4.649	5.358	5.279	4.619	4.141	5.011	4.935	4.286	4.272	3.794					
56 Ba	37.441	36.298	36.378	37.251	32.188	31.815	5.995		5.808	4.926	4.851	5.624	5.530	4.827	4.330	5.247	5.156	4.465	4.450	3.953					
57 La	38.925	37.714	37.801	38.723	33.436	33.034	6.267		6.073	5.143	5.061	5.891	5.788	5.041	4.524	5.483	5.383	4.650	4.633	4.124	0.851		0.854		
58 Ce	40.449	39.163	39.251	40.226	34.714	34.273	6.549		6.340	5.364	5.276	6.165	6.051	5.261	4.731	5.724	5.612	4.839	4.822	4.287	0.902		0.902		
59 Pr	41.998	40.646	40.741	41.767	36.020	35.544	6.846		6.615	5.591	5.497	6.443	6.321	5.488	4.935	5.968	5.849	5.033	5.013	4.452			0.949		
60 Nd	43.571	42.159	42.264	43.327	37.355	36.841	7.126		6.900	5.828	5.721	6.722	6.601	5.721	5.145	6.208	6.088	5.229	5.207	4.632	1.004		0.996		
61 Pm	45.207	43.705	43.818	44.929	38.718	38.171	7.448		7.485	6.070	6.195	7.018	6.891	5.960	5.588	6.466	6.338	5.432	5.407	4.994	1.108		1.100		
62 Sm	46.835	45.281	45.405	46.566	40.111	39.516	7.737		7.795	6.317	6.438	7.312	7.177	6.205	5.816	6.717	6.586	5.635	5.607	5.176			1.153		
63 Eu	48.515	46.896	47.030	48.248	41.535	40.895	8.069			6.570		7.624	7.479	6.455		6.983	6.842	5.845	5.816						
64 Gd	50.240	48.547	48.688	49.952	42.989	42.302	8.376		8.104	6.830	6.686	7.931	7.784	6.712	6.049	7.243	7.102	6.056	6.024	5.361	1.221		1.209		
65 Tb	51.996	50.221	50.374	51.715	44.474	43.737	8.708		8.422	7.095	6.939	8.252	8.100	6.977	6.283	7.515	7.365	6.272	6.237	5.546	1.280		1.266		
66 Dy	53.789	51.949	52.110	53.500	45.991	45.200	9.083		8.752	7.369	7.203	8.621	8.417	7.246	6.533	7.850	7.634	6.494	6.457	5.742			1.325		
67 Ho	55.615	53.702	53.868	55.315	47.539	46.692	9.395		9.086	7.650	7.470	8.919	8.746	7.524	6.787	8.071	7.910	6.719	6.679	5.942	1.390		1.383		
68 Er	57.483	55.485	55.672	57.204	49.119	48.213	9.776		9.429	7.938	7.744	9.263	9.087	7.809	7.057	8.364	8.188	6.947	6.904	6.152			1.443		
69 Tm	59.390	57.293	57.506	59.085	50.733	49.764	10.116		9.778	8.229	8.024	9.618	9.424	8.100	7.308	8.648	8.467	7.179	7.132	6.341	1.515		1.503		

Energy Table

X-ray emission and absorption energies (keV)

K - SERIES

Z Element	K(ab)	Kβ₃	Kβ₁	Kβ₂	Kα₁	Kα₂
70 Yb	61.332	59.141	59.356	60.974	52.380	51.345
71 Lu	63.304	61.037	61.272	62.956	54.061	52.956
72 Hf	65.351	62.969	63.222	64.969	55.781	54.602
73 Ta	67.414	64.398	65.212	67.001	57.523	56.267
74 W	69.524	66.940	67.233	69.089	59.308	57.972
75 Re	71.662	68.983	69.298	71.219	61.130	59.708
76 Os	73.860	71.065	71.401	73.390	62.990	61.486
77 Ir	76.112	73.190	73.548	75.606	64.885	63.276
78 Pt	78.395	75.355	75.735	77.864	66.821	65.112
79 Au	80.723	77.567	77.971	80.172	68.792	66.978
80 Hg	83.103	79.809	80.240	82.530	70.807	68.883
81 Tl	85.528	82.104	82.562	84.933	72.859	70.820
82 Pb	88.006	84.436	84.922	87.351	74.956	72.792
83 Bi	90.527	86.819	87.328	89.846	77.095	74.802
84 Po	93.112	89.231	89.781	92.383	79.279	76.851
85 At	95.740	91.707	92.287	94.974	81.499	78.930
86 Rn	98.418	94.230	94.850	97.622	83.768	81.051
87 Fr	101.147	96.791	97.460	100.307	86.089	83.217
88 Ra	103.927	99.415	100.113	103.051	88.454	85.419
89 Ac	106.759	102.084	102.829	105.849	90.868	87.660
90 Th	109.649	104.813	105.591	108.699	93.334	89.938
91 Pa	112.581	107.576	108.409	111.605	95.852	92.271
92 U	115.603	110.387	111.281	114.587	98.422	94.649
93 Np	118.619		113.725	118.057	100.781	96.844
94 Pu	121.760	116.943		120.350	103.300	99.168
95 Am	124.876	120.350		123.960	105.949	101.607
96 Cm	128.088	122.733		126.490	108.737	104.166
97 Bk	131.357	126.490	127.794	130.484	111.676	106.862
98 Cf	134.683			133.290	114.778	109.699

L - SERIES

	L₁ - SERIES				L_II - SERIES				L_III - SERIES				
Z Element	L₁(ab)	Lγ₃	Lβ₃	Lβ₄	L_II(ab)	Lγ₁	Lβ₁	Lη	L_III(ab)	Lβ₂	Lα₁	Lα₂	Ll
70 Yb	10.486	10.141	8.535	8.312	9.978	9.778	8.400	7.579	8.943	8.757	7.414	7.366	6.544
71 Lu	10.867	10.509	8.845	8.605	10.345	10.142	8.708	7.856	9.241	9.047	7.654	7.604	6.752
72 Hf	11.264	10.889	9.162	8.904	10.739	10.514	9.021	8.138	9.561	9.346	7.898	7.843	6.958
73 Ta	11.680	11.276	9.486	9.211	11.139	10.893	9.342	8.427	9.881	9.650	8.145	8.086	7.172
74 W	12.098	11.672	9.817	9.524	11.542	11.284	9.671	8.723	10.204	9.960	8.396	8.334	7.386
75 Re	12.522	12.080	10.158	9.845	11.955	11.683	10.008	9.026	10.531	10.274	8.651	8.585	7.602
76 Os	12.965	12.498	10.509	10.174	12.583	12.093	10.354	9.335	10.869	10.597	8.910	8.840	7.821
77 Ir	13.424	12.922	10.866	10.509	12.824	12.510	10.706	9.649	11.215	10.919	9.174	9.098	8.040
78 Pt	13.892	13.359	11.233	10.852	13.273	12.940	11.069	9.973	11.564	11.249	9.441	9.360	8.267
79 Au	14.353	13.807	11.608	11.203	13.733	13.379	11.440	10.307	11.918	11.583	9.712	9.626	8.493
80 Hg	14.846	14.262	11.993	11.561	14.209	13.828	11.821	10.649	12.284	11.922	9.987	9.896	8.720
81 Tl	15.344	14.734	12.388	11.929	14.698	14.289	12.211	10.992	12.657	12.270	10.267	10.171	8.952
82 Pb	15.860	15.215	12.791	12.304	15.198	14.762	12.612	11.347	13.035	12.621	10.550	10.448	9.183
83 Bi	16.385	15.708	13.208	12.689	15.708	15.245	13.021	11.710	13.418	12.978	10.837	10.729	9.419
84 Po	16.935		13.635	13.083	16.244	15.741	13.445		13.817	13.338	11.129	11.014	9.662
85 At	17.490		14.065		16.784	16.249	13.874		14.215		11.425	11.303	
86 Rn	18.058		14.509	14.745	17.337	16.768	14.313		14.618		11.725	11.596	
87 Fr	18.638	18.354	14.973		17.904	17.300	14.768	13.661	15.028	14.448	12.029	11.893	
88 Ra	19.233		15.442		18.484	17.845	15.233		15.442	14.839	12.338	12.194	
89 Ac	19.842	19.503	15.929	15.640	19.078	18.405	15.710	14.507	15.865		12.650	12.499	10.620
90 Th	20.470		16.423		19.692	18.979	16.199		16.300	15.621	12.967	12.807	11.117
91 Pa	21.102	20.094	16.927	16.101	20.311	19.565	16.699	14.944	16.731	16.022	13.288	13.120	11.364
92 U	21.756	20.709	17.452	16.573	20.947	20.164	17.217	15.397	17.167	16.425	13.612	13.437	11.616
93 Np	22.417	21.336	17.986	17.058	21.596	20.781	17.747	15.874	17.614	16.837	13.942	13.757	11.887
94 Pu	23.095	21.979	18.537	17.553	22.263	21.414	18.291	16.330	18.053	17.252	14.276	14.082	12.122
95 Am	23.793		19.103	18.060	22.944	22.061	18.849		18.526	17.673	14.615	14.409	12.381
96 Cm	24.503				23.640	22.703	19.399		18.990	18.096	14.953	14.740	
97 Bk	25.230				24.352	23.389	19.961		19.461	18.529	15.304	15.080	
98 Cf	25.971				25.080	24.070	20.557		19.938	18.983	15.652	15.418	

M - SERIES

	M_IV - SERIES		M_V - SERIES		
Z Element	M_IV(ab)	Mβ	M_V(ab)	Mα₁	Mα₂
70 Yb	1.578	1.567			
71 Lu		1.631			
72 Hf	1.718	1.697			
73 Ta	1.793	1.765			
74 W	1.871	1.835	1.809	1.775	1.773
75 Re		1.906			
76 Os	2.116	1.978	2.041	1.980	1.975
77 Ir	2.202	2.053	2.122	2.050	2.046
78 Pt		2.127			
79 Au	2.291	2.204	2.206	2.123	2.118
80 Hg	2.385	2.282	2.295		
81 Tl	2.485	2.362	2.389	2.270	2.265
82 Pb	2.586	2.442	2.484	2.345	2.339
83 Bi	2.687	2.525	2.579	2.422	2.416
84 Po					
90 Th	3.491	3.145	3.332	2.996	2.986
91 Pa		3.239		3.082	3.072
92 U	3.728	3.336	3.552	3.170	3.159

Glossary

argillaceous: Pertaining to, largely composed of, or containing clay-size particles or clay minerals.

authigenic: Formed or generated in place; specifically said of rock constituents and minerals that have not been transported or that were derived locally where they are found, and of minerals that came into existence at the same time, or subsequently to, the formation of the rock of which they constitute a part. The term, as used, often refers to a mineral (such as quartz or feldspar) formed after deposition of the original sediment.

birefringence: The ability of crystals other than those of the isometric system to split a beam of ordinary light into two beams of unequal velocities; the difference between the greatest and the least indices of refraction of a crystal.

birefringent: Said of a crystal that displays birefringence; such a crystal has more than one index of refraction.

coccolith: A general term applied to various microscopic calcareous structural elements or button-like plates having many different shapes and averaging about 3 microns in diameter (some have diameters as large as 35 microns), constructed of minute calcite or aragonite crystals, and constituting the outer skeletal remains of a coccolithophore. Coccoliths are found in chalk and in deep-sea oozes of the temperate and tropical oceans, and were probably not common before the Jurassic.

coccosphere: The entire spherical or spheroidal test or skeleton of a coccolithophore composed of an aggregation of interlocking coccoliths that are external to or embedded within an outer gelatinous layer of the cell. A coccolithophore.

conchoidal: Said of a type of mineral or rock fracture that gives a smoothly curved surface. It is a characteristic habit of quartz and of obsidian.

crenate: Having the edge, margin, or crest cut into rounded scallops or shallow rounded notches.

dedolomitization: A process whereby, presumably during contact metamorphism at low pressure, part or all of the magnesium in a dolomite or dolomitic limestone is used for the formation of magnesium oxides, hydroxides, and silicates (e.g., brucite, forsterite) resulting in the enrichment in calcite (Teall, 1903). The term was originally used by Morlot (1847) for the replacement of dolomite by calcite during diagenesis or chemical weathering.

detrital: Pertaining to or formed from detritus; said especially of rocks, minerals, and sediments.

diagenesis [sed]: All the chemical, physical, and biologic changes, modifications, or transformations undergone by a sediment after its initial deposition (i.e., after it has reached its final resting place in the current cycle of erosion, transportation, and deposition), and during and after its lithification, exclusive of surficial alteration (weathering) and metamorphism.

dissolution: A space or cavity in or between rocks, formed by the solution of part of the rock material.

druse: A mineral surface covered with small projecting crystals; specifically the crust or coating of crystals lining a druse in a rock, such as sparry calcite filling pore spaces in a limestone.

equant: Said of a crystal, in an igneous or sedimentary rock, having the same (or nearly the same) diameters in all directions. Synonym: equidimensional.

equigranular: Said of a rock texture having crystals of the same, or nearly the same, size.

euhedral: Said of a crystal, in a sedimentary rock (such as a calcite crystal in a recrystallized dolomite), characterized by the presence of crystal faces. Said of the shape of a euhedral crystal.

feldspathic litharenite: A term used by McBride (1963, p. 667) for a litharenite containing appreciable feldspar; specifically a sandstone containing 10 to 50% feldspar, 25 to 90% fine-grained rock fragments, and 0 to 65% quartz, quartzite, and chert.

ferruginous: Pertaining to or containing iron (e.g., a sandstone that is cemented with iron oxide).

fibrous: Said of the habit of a mineral, and of the mineral itself (e.g., asbestos), that crystallizes in elongated thin, needle-like grains, or fibers.

filiform: capillary.

framboid: A microscopic aggregate of pyrite grains, often in spheroidal clusters. It was considered to be the result of coloidal processes but is now linked with the presence of organic materials; sulfide crystals fill chambers or cells in bacteria (Park and MacDiarmid, 1970, p. 133).

micrograph: A graphic recording.

micropore: A pore small enough to hold water against the pull of gravity and to inhibit the flow of water.

overgrowth: Secondary material deposited in optical and crystallographic continuity around a crystal grain of the same composition, as in the diagenetic process of secondary enlargement.

paragenesis: The sequential order of mineral formation. A characteristic association or occurrence of minerals.

paragenetic: Pertaining to paragenesis. Pertaining to the genetic relations of sediments in laterally continuous and equivalent facies.

paramorphism: The property of a mineral to change its internal structure without changing its external form or chemical composition. Such a mineral is called a paramorph.

pellet: A small, usually rounded aggregate of accretionary material, such as a lapillus or a fecal pellet; specifically a spherical to elliptical (commonly ovoid, sometimes irregularly shaped) homogeneous clast made up almost exclusively of clay-sized calcareous (micritic) material, devoid of internal structure, and contained in the body of a well-sorted carbonate rock. Folk (1959; 1962) suggested that the term apply to allochems less than 0.15 to 0.20 mm in diameter, the larger grains being referred to as intraclasts, although in some rocks it is impossible to draw a sharp division. Pellets appear to be mainly the feces of mollusks and worms; others include pseudo-ooliths and aggregates produced by gas bubbling, by algal "budding" phenomena, or by other intraformational reworking of lithified or semilithified carbonate mud. A small rounded aggregate (0.1 to 0.3 mm in diameter) of clay minerals and fine quartz found in some shales and clays, separated from a matrix of the same materials by a shell of organic material, and ascribed to the action of water currents (Allen and Nichols, 1945).

planar: Lying or arranged as a plane or in planes, usually implying more or less parallel planes, such as those of bedding or cleavage. It is a two-dimensional arrangement, in contrast to the one-dimensional linear arrangement.

porcellanite: A hard, dense, siliceous rock having the texture, dull luster, hardness, fracture, or general appearance of unglazed porcelain; it is less hard, dense, and vitreous than chert.

pore [geol] : A small to minute opening or passageway in a rock or soil; an interstice.

pressure solution: Solution (in a sedimentary rock) occurring preferentially at the contact surfaces of grains (crystals) where the external pressure exceeds the hydraulic pressure of the interstitial fluid. It results in enlargement of the contact surfaces and thereby reduces pore space and tightly welds the rock.

pressolved: Said of a sedimentary bed or rock in which the grains have undergone pressure solution; e.g., "pressolved quartzite" whose toughness and homogeneity is due to a tightly interlocked texture of quartz grains subjected to pressure solution. Term was introduced by Heald (1956, p. 22).

pseudomorph: A mineral whose outward crystal form is that of another mineral species; it has developed by alteration, substitution, incrustation, or paramorphism. A pseudomorph is described as being *after* the mineral whose outward form it has (e.g., quartz after fluorite; Dana, p. 206).

resorption: The act or process of reabsorption or readsorption; specifically the partial or complete refusion or solution, by and in a magma, of previously formed crystals or minerals with which it is not in equilibrium or, owing to changes of temperature, pressure (depth), or chemical composition, with which it has ceased to be in equilibrium.

rhomb [cryst]: An oblique, equilateral parallelogram; in crystallography, a rhombohedron.

rhombohedron: A trigonal crystal form that is a parallelepiped whose six identical faces are rhombs. It is characteristic of the hexagonal system.

secondary porosity: The porosity developed in a rock formation subsequent to its deposition or emplacement, either through natural processes of dissolution or stress distortion, or artificially through acidization or the mechanical injection of coarse sand.

sparry: Pertaining to, resembling, or consisting of spar; (e.g., sparry vein or sparry luster). Pertaining to sparite, especially in allusion to the relative clarity both in thin section and hand specimen of the calcite cement; abounding with sparite, such as sparry rock.

sparry calcite: Clean, coarse-grained calcite crystal; sparite.

spectrum: (pl. spectra) An array of intensity values ordered according to any physical parameter, e.g., energy spectrum, mass spectrum, velocity spectrum.

sucrosic: A synonym of saccharoidal. The term is commonly applied to idiotopic dolomite rock.

vermiform: Worm-like or having the form of a worm (e.g., vermiform problematica consisting of long, thin, and more or less cylindrical tubes).

vitroclastic: Pertaining to a pyroclastic rock structure characterized by crescentically or triangularly fragmented bits of glass.

References

Recommended References: Introduction and General

AGI, 1972, Glossary of geology, M. Gary, R. McAfee, Jr., and C.L. Wolf, eds.: Falls Church, Virginia, American Geological Institute, 805 p.

Allen, V.T., and R.L. Nichols, 1945, Clay-pellet conglomerates at Hobart Butte, Lane County, Oregon: Journal of Sedimentary Petrology, v. 15, p. 25-33.

Bassin, N.J., 1975, Suspended marine clay mineral identification by scanning electron microscopy and energy-dispersive X-ray analysis: Limnology and Oceanography, v. 20, p. 133-137.

Beck, H.M., 1977, Schematic drawing of SEM/EDX system: unpub., 1 p.

Buchanan, R., 1983, SEM examination of non-conducting specimens: American Laboratory, April, p. 56-61.

Dana, E.S., 1892, The system of mineralogy by J.D. Dana, 1837-1868, descriptive mineralogy: New York, John Wiley and Sons, 1,134 p.

Deer, W.A., R.A. Howie, and J. Zussman, 1962, Rock-forming minerals: New York, John Wiley and Sons, vols. 1-5.

Everhart, T.E., and T.L. Hayes, 1972, The scanning electron microscope: Scientific American, v. 226, no. 1, p. 54-69.

Folk, R.L., 1959, Practical petrographic classification of limestones: AAPG Bulletin, v. 43, p. 1-38.

—**1962,** Spectral subdivision of limestone types, in W.E. Ham, ed., Classification of carbonate rocks - a symposium: AAPG Memoir No. 1, 279 p.

Goldstein, J.I. and H. Yakowitz, 1978, Practical scanning electron microscopy: New York, Plenum Press, 582 p.

Honjo, S., 1978, The scanning electron microscope in marine science: Oceanus, v. 21, no. 3, p. 19-29.

Kramers, J.W., and B.A. Rottenfusser, 1980, Techniques for SEM and EDX characterization of oil sands: Scanning Electron Microscopy, v. 4, p. 97-102.

McBride, E.F., 1963, A classification of common sandstones: Journal of Sedimentary Petrology, v. 33, p. 664-669.

Millot, G., 1970, Geology of clays (translated by W.R. Forrand and H. Pacquet): New York, Springer Verlag, 429 p.

Morlot, A. von, 1947, Ueber Dolomit und seine kunstliche Darstellung aus Kalkstein: Naturwissenschaftliche Abhandlungen, gesammelt and durch Subscription hrsg. von Willhelm Haidinger, v. 1, p. 305-315.

Mumpton, F.A., and W.C. Ormsby, 1976, Morphology of zeolites in sedimentary rocks by scanning electron microscopy: Clays and Clay Minerals, v. 24, p. 1-23.

Park, C.F., Jr., and R.A. MacDiarmid, 1970, Ore deposits: San Francisco, W.H. Freeman Company, 522 p.

Postek, M.T., et al, 1980, Scanning electron microscopy - a student's handbook: Burlington, Vermont, Ladd Research Industries, Inc., 305 p.

Roberts, W.L., G.R. Rapp, Jr., and J. Weber, 1974, Encyclopedia of minerals: New York, Van Nostrand Reinhold Co., 693 p.

Smith, D.G.W. (ed.), 1976, Short course in microbeam techniques: Mineralogical Association of Canada, 186 p.

Teall, J.J.H., 1903, On dedolomitisation: Geological Magazine, v. 10, p. 513-514.

Wells, O.C., 1974, Scanning electron microscopy: New York, McGraw-Hill, 421 p.

Wilson, M.D., and E.D. Pittman, 1977, Authigenic clays in sandstone: recognition and influence on reservoir properties and paleoenvironmental analysis: Journal of Sedimentary Petrology, v. 47, no. 1, p. 3-31.

Recommended References: Applications

Almon, W.R., 1979, A geologic appreciation of shaly sands: 20th Annual Logging Symposium, Society of Professional Well Log Analysts, Paper WW, 14 p.

—**1981,** Depositional environment and diagenesis of Permian Rotliegendes Sandstones in the Dutch sector of the southern North Sea, in F.J. Longstaffe, ed., Short course in clays and the resource geologist: Mineralogical Association of Canada, p. 119-147.

—**and D.K. Davies, 1981,** Formation damage and the crystal chemistry of clays, in F.J. Longstaffe, ed., Short course in clays and the resource geologist: Mineralogical Association of Canada, p. 81-102.

—**and A.L. Schultz, 1979,** Electric log detection in diagenetically altered reservoirs and diagenetic traps: Gulf Coast Association of Geological Societies Transactions, v. 29, p. 1-10.

—**L.B. Fullerton, and D.K. Davies, 1976,** Pore space reduction in Cretaceous sandstones through chemical precipitation of clay minerals: Journal of Sedimentary Petrology, v. 46, p. 89-96.

Barnes, D.J., and M.B. Dusseault, 1982, The influence of diagenetic microfabric on oil sands behavior: Canadian Journal of Earth Sciences, v. 19, no. 4, p. 804-818.

Dann, M.W., D.B. Burnett, and L.M. Hall, 1982, Polymer performance in low permeability reservoirs: Society of Petroleum Engineers 10615, p. 201-207.

Davies, D.K., and W.R. Almon, 1977, Effects of sandstone composition and diagenesis on reservoir quality, Tertiary-Pleistocene, Gulf Coast region: Gulf Coast Association of Geological Societies Transactions, v. 27, p. 197.

Dengler, L.A., 1980, The microstructure of deformed graywacke sandstones: Livermore, California, Lawrence Livermore Laboratory, UCID-18638, 273 p.

Frank, J.R., S. Cluff, and J.M. Bauman, 1982, Painter reservoir, East Painter reservoir, and Clear Creek fields, Uinta County, Wyoming, in R.B. Powers, ed., Geologic studies of Cordilleran thrust belts: Denver, Rocky Mountain Association of Geologists, p. 601-611.

Hancock, N.J., 1978, An application of scanning electron microscopy in pilot water injection studies for oilfield development, in W.B. Whalley, ed., Scanning electron microscopy in the study of sediments: Norwich, England, Geological Abstracts, p. 61-70.

Hempkins, W.B., A. Timur, and R.M. Weinbrandt, 1971, Scanning electron microscope study of pore systems in rocks: Journal of Geophysical Research, v. 76, no. 20, p. 4932-4948.

Kieke, E.M., and D.J. Hartman, 1973, Scanning electron microscope application to formation evaluation: Gulf Coast Association of Geological Societies Transactions, v. 23, p. 60-67.

Kupperman, G.S., B.M. Ward, and G.L. Blank, 1982, Scanning electron microscopy and core analysis for "mini" CO_2 flooding operations in the Hospah Formation of the Miguel Creek field, McKinley County, New Mexico: Society of Petroleum Engineers Journal, v. 22, no. 6 (December), p. 797-804.

Lambert-Aikhionbare, D.O., 1982, Relationship between diagenesis and pore fluid chemistry in Niger Delta oil-bearing sands: Journal Petroleum Geology, v. 4, no. 3, p. 287-298.

Lindquist, S.J., 1983, Nugget Formation reservoir characteristics affecting production in the overthrust belt of southwestern Wyoming: Journal of Petroleum Technology, July, p. 1355-1365.

Link, M.H., and J.E. Welton, 1982, Sedimentology and reservoir potential of Matilija Sandstone - an Eocene sand-rich deepsea fan and shallow marine complex, California: AAPG Bulletin, v. 66, no. 10, p. 1514-1534.

Longstaffe, F.J., ed., **1981**, Short course in clays and the resource geologist: Mineralogical Association of Canada, 199 p.

McCoy, J.T., **1977**, Petrophysical evaluation of the Bluesky sand, Bassett area, Alberta: Bulletin of Canadian Petroleum Geology, v. 25, p. 378-395.

McLaughlin, H.C., Sr., et al, **1977**, Clay stabilizing agent can correct formation damage: World Oil, v., p. 58.

Neasham, J.W., **1977**, Applications of scanning electron microscopy to characterization of hydrocarbon-bearing rocks: Scanning Electron Microscopy, v. 10, p. 101-108.

Nydegger, G.L., D.D. Rice, and C.A. Brown, **1980**, Analysis of shallow gas development from low permeability reservoirs of lower Cretaceous age, Bowdoin Dome area: Journal of Petroleum Technology, v. 32, no. 12, p. 2111-2120.

Pittman, E.D., **1979**, Porosity, diagenesis, and prouctive capability of sandstone reservoirs, *in* P.A. Scholle and P.R. Schluger, eds., Aspects of diagenesis: SEPM Special Publication 26, p. 159-173.

—and R.W. Duschatko, **1970**, Use of pore casts and scanning electron microscope to study pore geometry: Journal of Sedimentary Petrology, v. 40, p. 1153-1157.

—and J.B. Thomas, **1978**, Some applications of scanning electron microscopy to the study of reservoir rocks: Society of Petroleum Engineers 7550, 4 p.

Pye, K., and D. Krinsley, **1983**, Mudrocks examined by backscatter electron microscopy: Nature, v. 301, p. 412-413.

Sarkisyan, S.G., **1971**, Application of the SEM in the investigation of oil and gas reservoir rocks: Journal of Sedimentary Petrology, v. 41, p. 289-292.

Sassen, R., **1980**, Biodegradation of crude oil and mineral deposition in a shallow Gulf Coast salt dome: Organic Geochemistry, v. 2, p. 153-166.

Scholle, P.A., and P.R. Schluger, **1979**, Aspects of diagenesis: SEPM Special Publication 26, 443 p.

Schrank, J.A., and E. Hunt, **1980**, Improved reservoir evaluation with the SEM: Scanning Electron Microscopy, v. 1, p. 573-578.

Schultz, A.L., **1979**, Electric log evidence for hydrocarbon production and trapping in sandstones possessing diagenetic clay minerals: Bulletin South Texas Geological Society, v. 19, no. 8, p. 24-28.

Simon, D.E., F.W. Kaul, and J.N. Culberston, **1979**, Anadarko basin Morrow-Springer sandstone simulation study: Journal of Petroleum Technology, June, p. 683-689.

Smart, P., and N.K. Tovey, **1981**, Electron microscopy of soils and sediments - examples: Oxford, England, Clarendon Press, 178 p.

Somerton, W.H., and C.J. Radke, **1983**, Role of clays in the enhanced recovery of petroleum from some California sands: Journal of Petroleum Technology, March, p. 643-654.

Swanson, B.F., **1977**, Visualizing pores and non-wetting phase in porous rock: Society of Petroleum Engineers 6857, 10 p.

Thomas, J.B., **1978**, Diagenetic sequences in low-permeability argillaceous sandstones: Journal of the Geological Society of London, v. 135, p. 93-99.

—**1981**, Classification and diagenesis of clay minerals in tight gas sandstones: case studies in which clay mineral properties are crucial to drilling fluid selection, formation evaluation, and completion techniques, *in* F.J. Longstaffe, ed.; Short course in clays and the resource geologist: Mineralogical Association of Canada, p. 104-118.

Thomas, M. and B. Miller, **1980**, Diagenesis and rock-fluid interactions in the Cadotte Member from a well in northeastern British Columbia: Bulletin of Canadian Petroleum Geology, v. 28, p. 173-199.

Wardlaw, N.C., and J.P. Cassan, **1979**, Oil recovery efficiency and the rock-pore properties of some sandstone reservoirs: Bulletin of Canadian Petroleum Geology, v. 27, no. 2, p. 117-138.

Weinbrandt, R.M., and I. Fatt, **1969**, A scanning electron microscope study of the pore structure of sandstone: Journal of Petroleum Technology, May, p. 543-548.

Whalley, W.B., ed., **1979**, Scanning electron microscopy in the study of sediments: Norwich, England, Geological Abstracts, 414 p.

Recommended References: Silicates (Silica)

Douglas, L.A., and D.W. Platt, **1977**, Surface morphology of quartz and age of soils: Soils Science Society of America, v. 41, p. 641-645.

Heald, M.T., **1956**, Cementation of Simpson and St. Peter sandstones in parts of Oklahoma, Arkansas, and Missouri: Journal of Geology, v. 64, p. 16-30.

Hill, P.J., and J.D. Collen, **1978**, The Kapuni sandstones from Inglewood No. 1 well, Taranaki - petrology and effect of diagenesis on reservoir characteristics: New Zealand Journal of Geology and Geophysics, v. 21, p. 215-228.

Holland, M.T., **1982**, Reservoir property implications of pore geometry modification accompanying sand diagenesis: Anahuac Formation, Louisiana: 57th Annual Fall Conference, Society of Petroleum Engineers, Paper SPE 10991, 6 p.

Isaacs, C.M., **1981**, Porosity reduction during diagenesis of the Monterey Formation, Santa Barbara coastal area, California, *in* R.E. Garrison et al, eds., The Monterey Formation and related siliceous rocks of California: Pacific Section SEPM, p. 257-271.

Krinsley, D. and J. Donahue, **1968**, Environmental interpretation of sand grain surface texture by electron microscopy: Geological Society of America Bulletin, v. 79, p. 743-748.

—and J.C. Doornkamp, **1973**, Atlas of quartz sand surface textures: Cambridge, England, Cambridge University Press, 91 p.

—and S. Margolis, **1969**, A study of quartz sand grain surfaces with SEM: New York Academy of Sciences Transactions, v. 31, p. 457-477.

Lancelot, Y., **1973**, Chert and silica diagenesis in sediments from the central Pacific, *in* Initial reports of the deep sea drilling project: Washington, D.C., U.S. Government Printing Office, v. 17, p. 377-405.

Martin, K.R., and N.J. Hamilton, **1981**, Diagenesis and reservoir quality, Toolachee Formation, Cooper basin: APEA Journal, v. 21, p. 143-154.

Marzolf, J.E., **1976**, Sand-grain frosting and quartz overgrowth examined by scanning electron microscopy - the Navajo Sandstone, (Jurassic?), Utah: Journal of Sedimentary Petrology, v. 46, p. 906-912.

Oehler, J.H., **1975**, Origin and distribution of silica lepispheres in porcelanite from the Monterey Formation of California: Journal of Sedimentary Petrology, v. 45, p. 252-257.

Pittman, E.D., **1972**, Diagenesis of quartz in sandstones as revealed by scanning electron microscopy: Journal of Sedimentary Petrology, v. 42, no. 3, p. 507-519.

Riezebos, P.A., **1974**, Scanning electron microscopical observations on weakly cemented Miocene sands: Geologie en Mijnbouw, v. 53, p. 109-122.

Scholle, P.A., **1979**, A color illustrated guide to constituents, textures, cements, and porosities of sandstones and associated rocks: AAPG Memoir 28, 201 p.

Stein, C.L., **1982**, Silica recrystallization in petrified wood: Journal of Sedimentary Petrology, v. 52, p. 1277-1284.

Subramanian, V., 1975, Origin on surface pits on quartz as revealed by scanning electron microscopy: Journal of Sedimentary Petrology, v. 45, p. 530-534.

Tankard, A.J., and D.H. Krinsley, 1977, Diagenetic surface textures on quartz grains - an application of SEM: Transactions of the Geological Society of South Africa, v. 77, p. 285-287.

Thomas, J.B., 1978, Diagenetic sequences in low-permeability argillaceous sandstones: Journal of the Geological Society of London, v. 135, p. 93-99.

Thomson, A., 1978, Petrography and diagenesis of the Hosston Sandstone reservoirs at Bassfield, Jefferson Davis County, Mississippi: Gulf Coast Association of Geological Societies Transactions, v. 28, p. 651-664.

Waugh, B., 1970, Form of quartz overgrowths in the Penrith sandstone (lower Permian) of northwest England as revealed by SEM: Sedimentology, v. 14, p. 309-320.

—1978, Diagenesis in continental redbeds as revealed by scanning electron microscopy - a review, in W.B. Walley, ed., Scanning electron microscopy in the study of sediments: Norwich, England, Geological Abstracts, p. 329-346.

Whalley, W.B., 1978, Earth surface diagenesis of an orthoquartzite - scanning electron microscope examination of sarsen stones from southern England and silcretes from Australia, in W.B. Whalley, ed., Scanning electron microscopy in the study of sediments: Norwich, England, Geological Abstracts, p. 383-398.

Wilson, P., 1978, Quartz overgrowths from the Millstone Grit Sandstones (Namurian) of the southern Pennines as revealed by scanning electron microscopy: Proceedings of the Yorkshire Geological Society, v. 42, p. 289-295.

Recommended References: Silicates (Feldspars)

Ali, A.D., and P. Turner, 1982, Authigenic K-feldspar in the Bromsgrove Sandstone Formation (Triassic) of central England: Journal of Sedimentary Petrology, v. 52, p. 187-197.

Boles, J.R., 1982, Active albitization of plagioclase, Gulf Coast Tertiary: American Journal of Science, v. 282, p. 165-180.

Kastner, M., and R. Siever, 1979, Low temperature feldspars in sedimentary rocks: American Journal of Science, v. 279, p. 435-479.

Odom, I.E., T.N. Willand, and R.J. Lassin, 1979, Paragenesis of diagenetic minerals in the St. Peter Sandstone (Ordovician), Wisconsin and Illinois, in P.A. Scholle and P.R. Schluger, eds., Aspects of diagenesis: SEPM Special Publication 26, p. 425-443.

Scholle, P.A., 1979, A color illustrated guide to constituents, textures, cements, and porosites of sandstones and associated rocks: AAPG Memoir 28, 201 p.

Stablein, N.K., and E.C. Dapples, 1977, Feldspars of the Tunnel City Group (Cambrian), western Wisconsin: Journal of Sedimentary Petrology, v. 47, p. 1512-1538.

Waugh, B., 1978, Authigenic K-feldspar in British Permo-Triassic sandstones: Journal of the Geological Society of London, v. 135, p. 51-56.

Recommended References: Silicates (Clays)

Al-Gailani, M.B., 1981, Authigenic mineralization at unconformities; implications for reservoir characteristics: Sedimentary Geology, v. 29, p. 89-115.

Almon, W.R., L.B. Fullerton, and D.K. Davies, 1976, Pore space reduction in Cretaceous sandstones through chemical precipitation of clay minerals: Journal of Sedimentary Petrology, v. 46, p. 89-96.

Bjorlykke, K., A. Elverhol, and A.O. Malm, 1979, Diagenesis in Mesozoic sandstones from Spitsbergen and the North Sea - a comparison: Geologie Rundschau, v. 68, p. 1152-1171.

Bohor, B.F., and R.E. Hughes, 1971, Scanning electron microscopy of clays and clay minerals: Clays and Clay Minerals, v. 19, no. 1, p. 49-54.

Borst, R.L., and R.Q. Gregg, 1969, Authigenic mineral growth as revealed by the scanning electron microscope: Journal of Sedimentary Petrology, v. 39, p. 1596-1597.

Borst, R.L., and W.D. Keller, 1969, Scanning electron micrographs of API reference clay minerals and other selected samples: International Clay Conference, v. 1, p. 871-901.

Colter, V.S., and J. Ebbern, 1979, SEM studies of Triassic reservoir sandstones from the Morecambe field, Irish Sea, U.K.: Scanning Electron Microscopy, v. 1, p. 531-538.

Dypvik, H., and J. Vollset, 1979, Petrology and diagenesis of Jurassic sandstone from Norwegian Danish basin; North Sea: AAPG Bulletin, v. 63, no. 2, p. 182-193.

Edwards, M.D., 1979, Sandstone in lower Cretaceous Helvetiafjellet Formation, Svalbard: Bearing on reservoir potential of Barents shelf: AAPG Bulletin, v. 63, no. 12, p. 2193-2203.

Eswaran, H., 1972, Morphology of allophane, imogolite and halloysite: Clays and Clay Minerals, v. 9, p. 281-284.

Foscolos, A.E., G.E. Reinson, and T.G. Powell, 1982, Controls on clay-mineral authigenesis in the Viking sandstone, central Alberta - 1. Shallow depths: Canadian Mineralogists, v. 20, p. 141-150.

Frank, J.R., S. Cluff, and J.M. Bauman, 1982, Painter reservoir, East Painter reservoir and Clear Creek fields, Uinta County, Wyoming, in R.B. Powers, ed., Geologic studies in Cordilleran thrust belts: Rocky Mountain Association of Geologists, p. 601-611.

Guven, N., W.E. Hower, and D.K. Davies, 1980, Nature of authigenic illites in sandstone reservoirs: Journal of Sedimentary Petrology, v. 50, p. 761-766.

Hammond, C., et al, eds., 1980, Selected papers from the symposium on the microscopy of clays and soils: Journal of Microscopy, v. 120, pt. 3, p. 235-366.

Hancock, N.J., 1978, Possible causes of Rotliegendes sandstone diagenesis in northern West Germany: Journal of the Geological Society of London, v. 135, p. 35-40.

—and A.M. Taylor, 1978, Clay mineral diagenesis and oil migration in the Middle Jurassic Brent Sand Formation: Journal of the Geological Society of London, v. 135, p. 69-72.

Hansley, P.L., and R.C. Johnson, 1980, Mineralogy and diagenesis of low-permeability sandstones of Late Cretaceous age, Piceance Creek basin, northwestern Colorado: Mountain Geologist, v. 17, p. 88-129.

Hayes, J.B., 1970, Polytypism of chlorite in sedimentary rocks: Clays and Clay Minerals, v. 18, p. 285-306.

Hurst, A., and H. Irwin, 1982, Geological modelling of clay diagenesis in sandstone: Clay Minerals, v. 17, p. 5-22.

Hutcheon, I., A. Oldershaw, and E.D. Ghent, 1980, Diagenesis of Cretaceous sandstones of the Kootenay Formation at Elk Valley (southeastern British Columbia) and Mt. Allan (southwestern Alberta): Geochimica et Cosmochimica Acta, v. 44, p. 1425-1435.

Iwuagwu, C.J., and J.F. Lerbekmo, 1981, The role of authigenic clays in some reservoir characteristics of the basal Belly River sandstone, Pembina field, Alberta: Bulletin of Canadian Petroleum Geology, v. 29, p. 479-491.

Jeans, C.V., et al, 1982, Volcanic clays in the Cretaceous of southern England and northern Ireland: Clay Minerals, v. 17, p. 105-156.

Keighin, C.W., 1979, Influence of diagenetic reactions on reservoir properties of the Neslen, Farrer, and Tuscher Formations, Uinta Basin, Utah: Society of Petroleum Engineers Symposium on low permeability gas reservoirs, paper SPE 7919, p. 77-80.

—1980, Evaluation of pore geometry of some low-permeability sandstones, Uinta basin, Utah: 55th Annual Fall Conference, Society of Petroleum Engineers, Paper SPE 9251, 4 p.

Keller, W.D., 1976a, Scan electron micrographs of kaolins collected from diverse environments of origin, Parts 1 and 2: Clays and Clay Minerals, v. 24, p. 107-117.

—1976b, Scan electron micrographs of kaolins collected from diverse origins, part 3: influence of parent material on flint clays and flint-like clays: Clays and Clay Minerals, v. 24, p. 262-264.

—1977, Scan electron micrographs of kaolins collected from diverse origins, Part IV: Clays and Clay Minerals, v. 25, p. 311-345.

—1978a, Classification of kaolins exemplified by their textures in SEM: Clays and Clay Minerals, v. 26, no. 1, p. 1-20.

—1978b, Kaolinization of feldspar as displayed in SEM micrographs: Geology, v. 6, p. 184-188.

—1982, Kaolin - a most diverse rock in genesis, texture, physical properties, and uses: Geological Society of America Bulletin, v. 93, p. 27-36.

—et al, 1980, Kaolin from the original Kauling (Gaoling) Mine locality, Kiangsi Province, China: Clays and Clay Minerals, v. 28, no. 2, p. 97-104.

Kerr, P.F., 1950, Analytical data on reference clay minerals (Preliminary Report No. 7: Reference Clay Minerals): American Petroleum Institute, New York, Columbia University, 160 p.

McConchie, D.M., and D.W. Lewis, 1978, Authigenic, perigenic, and allogenic glauconites from the Castle Hill basin, North Canterbury, New Zealand: New Zealand Journal of Geology and Geophysics, v. 21, p. 199-214.

McHardy, W.J., M.J. Wilson, and J.M. Tait, 1982, Electron microscope and X-ray diffraction studies of filamentous illitic clay from sandstones of the Magnus field: Clay Minerals, v. 17, p. 23-39.

Morris, K.A., and C.M. Shepperd, 1982, The role of clay minerals in influencing porosity and permeability characteristics in the Bridport Sands of Wytch Farm, Dorset: Clay Minerals, v. 17, p. 41-54.

Morris, R.C., K.E. Proctor, and M.R. Koch, 1979, Petrology and diagenesis of deep-water sandstones, Ouachita Mountains, Arkansas and Oklahoma, in P.A. Scholle and P.R. Schluger, eds., Aspects of diagenesis: SEPM Special Publication 26, p. 263-279.

Neasham, J.W., 1977, The morphology of dispersed clay in sandstone reservoirs and its effect on sandstone shaliness, pore space and fluid flow properties: Society of Petroleum Engineers, Paper SPE 6858, 3 p.

Odom, I.E., 1976, Microstructure, mineralogy and chemistry of Cambrian glauconite pellets and glauconite, central U.S.A.: Clays and Clay Minerals, v. 24, p. 232-238.

—T.N. Willand, and R.J. Lassin, 1979, Paragenesis of diagenetic minerals in the St. Peter Sandstone (Ordovician), Wisconsin and Illinois, in P.A. Scholle and P.R. Schulger, eds., Aspects of diagenesis: SEPM Special Publication 26, p. 425-443.

Pittman, E.D., 1979, Porosity, diagenesis and productive capability of sandstone reservoirs, in P.A. Scholle and P.R. Schluger, eds., Aspects of diagenesis: SEPM Special Publication 26, p. 159-173.

Pollastro, R.M., 1981, Authigenic kaolinite and associated pyrite in chalk of the Cretaceous Niobrara Formation, Eastern Colorado: Journal of Sedimentary Petrology, v. 51, no. 2, p. 553-562.

Rossel, N.C., 1982, Clay mineral diagenesis in Rotliegand aeolin sandstones of the southern North Sea: Clay Minerals, v. 17, p. 69-77.

Sarkisyan, S.G., 1971, Application of the SEM in the investigation of oil and gas reservoir rocks: Journal of Sedimentary Petrology, v. 41, p. 289-292.

Scholle, P.A., 1979, A color illustrated guide to constituents, textures, cements and porosities of sandstones and associated rocks: AAPG Memoir 28, 201 p.

Schultz, A.L., 1979, Electric log evidence for hydrocarbon production and trapping in sandstones possessing diagenetic clay minerals: Bulletin of South Texas Geological Society, v. 19, no. 8, p. 24-28.

Sedimentology Research Group, 1981, The effects of in situ steam injection on Cold Lake oil sands: Bulletin of Canadian Petroleum Geology, v. 29, p. 447-478.

Seeman, U., 1979, Diagenetically formed interstitial clay minerals as a factor in Rotliegand sandstone reservoir quality in the Dutch Sector of the North Sea: Journal of Petroleum Geology, v. 1, no. 3, p. 55-62.

—1982, Depositional facies, diagenetic clay minerals and reservoir quality of Rotliegand sediments in the Southern Permian basin (North Sea) — a review: Clay Minerals, v. 17, p. 55-67.

Sommer, F., 1975, Histoire diagenetique d'une serie greseuse de mer du Nord. Datation de l'introduction des hydrocarbones: Revue de l'institut Francais de Petrole, v. 30, no. 5, p. 729-742.

—1978, Diagenesis of Jurassic sandstones in the Viking Graben: Journal of the Geological Society of London, v. 135, p. 63-67.

Stalder, P.J., 1973, Influence of crystallographic habit and aggregate structure of authigenic clay minerals on sandstone permeability: Geologie en Mijnbouw, v. 52, p. 217-220.

Thomson, A., 1982, Preservation of porosity in the Deep Woodbine/Tuscaloosa Trend, Louisiana: Journal of Petroleum Technology, v. 34, no. 5, p. 1156-1162.

Tompkins, R.E., 1981, Scanning electron microscopy of a regular chlorite/smectite (corrensite) from a hydrocarbon reservoir sandstone: Clays and Clay Minerals, v. 29, no. 3, p. 233-235.

Walker, T.R., B. Waugh, and A.J. Crone, 1978, Diagenesis of first-cycle desert alluvium of Cenozoic age, southwestern U.S. and northwestern Mexico: Geological Society of America Bulletin, v. 89, p. 19-32.

Webb, J.E., 1974, Relation of oil migration to secondary clay cementation, Cretaceous sandstones, Wyoming: AAPG Bulletin, v. 58, p. 2245-2249.

Welton, J.E., and M.H. Link, 1982, Diagenesis of sandstones from Miocene-Pliocene Ridge basin, Southern California, in Geologic history of the Ridge basin, southern California: Pacific Section, SEPM, p. 181-190.

Wilson, M.D., 1982, Origins of clays controlling permeability in tight gas sands: Journal of Petroleum Technology, v. 34, no. 12, p. 2871-2876.

Wilson, M.D., and E.D. Pittman, 1977, Authigenic clays in sandstone: recognition and influence on reservoir properties and paleoenvironmental analysis: Journal of Sedimentary Petrology, v. 47, no. 1, p. 3-31.

Recommended References: Silicates (Zeolites)

Bernoulli, D., R.E. Garrison, and F. Melieres, 1978, Phillipsite cementation in a foraminiferal sandstone at Hole 373A and "The case of the violated foram," in Initial reports deep sea drilling project, v. 42: Washington, D.C., U.S. Government Printing Office, p. 478-482.

Davies, D.K., et al, 1979, Deposition and diagenesis of Tertiary - Holocene volcaniclastics, in P.A. Scholle and P.R. Schluger, eds., Aspects of diagenesis: SEPM Special Publication 26, p. 281-306.

Gude, A.J., and R.A. Sheppard, 1981, Woolly erionite from the Reese River zeolite deposit, Lander county, Nevada, and its relationship to other erionites: Clays and Clay Minerals, v. 29, p. 378-384.

McCulloh, T.H., et al, 1981, Precipitation of Laumontite with quartz, thenardite, and gypsum at Sespe Hot Springs, western Transverse Ranges, California: Clays and Clay Minerals, v. 29, p. 353-364.

Moncure, G.K., R.C. Surdam, and H.L. McKague, 1981, Zeolite diagenesis below Pahute Mesa, Nevada Test Site: Clays and Clay Minerals, v. 29, p. 385-396.

Mumpton, F.A., and W.C. Ormsby, 1976, Morphology of zeolites in sedimentary rocks by scanning electron microscopy: Clays and Clay Minerals, v. 24, p. 1-23.

Richmann, D.L., et al, 1980, Mineralogy, diagenesis, and porosity in Vicksburg sandstones, McAllen Ranch field, Hidalgo County, Texas: Gulf Coast Association of Geological Societies Transactions, v. 30, p. 473-481.

Sommer, F., 1978, Diagenesis of Jurassic sandstones in the Viking Graben: Journal of the Geological Society of London, v. 135, p. 63-67.

Stanley, K.O., and L.V. Benson, 1979, Early diagenesis of high plains Tertiary vitric and arkosic sandstone, Wyoming and Nebraska, in P.A. Scholle and P.R. Schluger, eds., Aspects of diagenesis: SEPM Special Publication 26, p. 401-423.

Surdam, R.C., and J.R. Boles, 1979, Diagenesis of volcanic sandstones, in P.A. Scholle and P.R. Schluger, eds., Aspects of diagenesis: SEPM Special Publication 26, p. 227-242.

Taylor, M.W., and R.C. Surdam, 1971, Zeolitic reactions in the tuffaceous sediments at Teels Marsh, Nevada: Clays and Clay Minerals, v. 29, p. 341-352.

Wise, W.S., and R.W. Tschernich, 1978, Habits, crystal forms and composition of Thomsonite: Canadian Mineralogist, v. 16, p. 487-493.

Recommended References: Silicates (Micas)

Sedimentology Research Group, 1981, The effects of in situ steam injection on Cold Lake oil sands: Bulletin of Canadian Petroleum Geology, v. 29, p. 447-478.

Sommer, F., 1978, Diagenesis of Jurassic sandstones in the Viking Graben: Journal of the Geological Society of London, v. 135, p. 63-67.

Recommended References: Silicates (Amphiboles, Pyroxenes and Others)

Lin, I.J., V. Rohrlich, and A. Slatkine, 1974, Surface microtextures of heavy minerals from the Mediterranean coast of Israel: Journal of Sedimentary Petrology, v. 44, p. 1281-1295.

Rahmani, R.A., 1973, Grain surface etching features of some heavy minerals: Journal of Sedimentary Petrology, v. 43, p. 882-888.

Setlow, L.W., and R.P. Karpovich, 1972, "Glacial" micro-texture on quartz and heavy mineral sand grains from the littoral environments: Journal of Sedimentary Petrology, v. 42, p. 864-875.

Simpson, G.S., 1976, Evidence of overgrowth on, and solution of, detrital garnets: Journal of Sedimentary Petrology, v. 46, p. 689-693.

Walker, T.R., B. Waugh, and A.J. Crone, 1978, Diagenesis of first-cycle desert alluvium of Cenozoic age, southwestern U.S. and northwestern Mexico: Geological Society of America Bulletin, v. 89, p. 19-32.

Waugh, B., 1978, Diagenesis in continental red beds as revealed by scanning electron microscopy - a review, in W.B. Whalley, ed., Scanning electron microscopy in the study of sediments: Geological Abstracts, Norwich, England, p. 329-346.

Recommended References: Carbonates

Al-Shaieb, Z., and J.W. Shelton, 1978, Secondary ferroan dolomite rhombs in oil reservoirs, Chadra sands, Gialo field, Libya: AAPG Bulletin, v. 62, p. 463-468.

Armstrong, A.K., P.D. Snavely, Jr., and W.O. Addicott, 1980, Porosity evaluation of Upper Miocene reefs, Almeria Province, southern Spain: AAPG Bulletin, v. 64, no. 2, p. 188-208.

Blanche, J.B., and J.H. McD Whitaker, 1978, Diagenesis of part of the Brent Sand Formation (Middle Jurassic) of the northern North Sea basin: Journal of the Geological Society of London, v. 135, p. 73-82.

Bricker, O.P., ed., 1971, Carbonate cements: Baltimore, Johns Hopkins Press, 376 p.

Ginsburg, R.N., and J.H. Schroeder, 1973, Growth and submarine fossilization of algal cup reefs, Bermuda: Sedimentology, v. 20, p. 575-614.

—and N.P. James, 1976, Submarine botryoidal aragonite in Holocene reef limestones, Belize: Geology, v. 4, p. 431-436.

James, N.S., et al, 1976, Facies and fabric specificity of early subsea cements in shallow Belize (British Honduras) reefs: Journal of Sedimentary Petrology, v. 46, no. 3, p. 523-544.

Longman, M.W., 1980, Carbonate diagenetic textures from near surface diagenetic environments: AAPG Bulletin, v. 64, no. 4, p. 461-487.

—and P.A. Mench, 1978, Diagenesis of Cretaceous limestones in the Edwards aquifer system of South-Central Texas - A scanning electron microscope study: Sedimentary Geology, v. 21, p. 241-276.

MacIntyre, I.G., 1977, Redistribution of submarine cements in a modern Caribbean fringing reef, Galeta Point, Panama: Journal of Sedimentary Petrology, v. 47, no. 2, p. 503-516.

Mou, D.C., and R.L. Brenner, 1982, Control of reservoir properties of Tensleep Sandstone by depositional and diagenetic facies and Lost Soldier field, Wyoming: Journal of Sedimentary Petrology, v. 52, p. 367-381.

Naiman, E.R., A. Bein, and R.L. Folk, 1983, Complex polyhedral crystals of limpid dolomite associated with halite, Upper Clear Fork and Glorietta Formations, Texas: Journal of Sedimentary Petrology, v. 53, p. 549-555.

Scholle, P.A., 1977, Chalk diagenesis and its relation to petroleum exploration: Oil from chalk, a modern miracle?: AAPG Bulletin, v. 61, p. 982-1009.

—1978, A color illustrated guide to carbonate rocks constitutents, textures, cements and porosities: AAPG Memoir 27, 254 p.

—1981, Porosity predication in shallow vs. deepwater limestones: Journal of Petroleum Technology, v. 33, p. 2236-2242.

Schroder, J.H., 1972, Fabrics and sequences of marine carbonate cements in Holocene Bermuda cup reefs: Geologie Rundschau, v. 61, p. 708-730.

Thomas, M.B., and T.A. Oliver, 1979, Depth-porosity relationships in the Viking and Cardium Formations of central Alberta: Bulletin of Canadian Petroleum Geology, v. 27, p. 209-228.

Welton, J.E., and M.H. Link, 1982, Diagenesis of sandstones from Miocene-Pliocene Ridge basin, southern California, in Geologic history of the Ridge basin, southern California: Pacific Section, SEPM, p. 181-190.

Recommended References: Phosphates

Hearn, P.P., D.L. Parkhurst, and E. Callender, 1983, Authigenic vivianite in Potomac river sediments - control by ferric oxy-hydroxides: Journal of Sedimentary Petrology, v. 53, p. 165-177.

Recommended References: Halides

Eswaran, H., G. Stoops, and A. Abtahi, 1980, SEM morphology of halite (NaCl) in soils: Journal of Microscopy, v. 120, p. 343-352.

Recommended References: Sulfides

Elverhoi, A., 1977, Origin of framboidal pyrite in clayey Holocene sediments and in Jurassic black shale in the northwestern part of the Barents Sea: Sedimentology, v. 24, p. 591-595.

Pollastro, R.M., 1981, Authigenic kaolinite and associated pyrite in chalk of the Cretaceous Niobrara Formation, eastern Colorado: Journal of Sedimentary Petrology, v. 51, p. 553-562.

Recommended References: Sulfates

Glennie, K.W., G.C. Mudd, and P.J.C. Nagtegaal, 1978, Depositional environment and diagenesis of Permian Rotliegendes Sandstone in Leman Bank and Sole Pit areas of the U.K., southern North Sea: Journal of the Geological Society of London, v. 135, p. 25-34.

Kessler, L.G., II, 1978, Diagenetic sequence in ancient sandstones deposited under desert climatic conditions: Journal of the Geological Society of London, v. 135, p. 41-49.

Mankiewicz, D., and J.R. Steidtmann, 1979, Depositional environments and diagenesis of the Tensleep Sandstone, eastern Big Horn basin, Wyoming, in P.A. Scholle and P.R. Schluger, eds., Aspects of diagenesis: SEPM Special Publication 26, p. 319-336.

Recommended References: Oxides

Eswaran, H., and N. Daud, 1980, Scanning electron microscopy evaluation of soils from Malayasia: Soils Science Society of America, v. 44, p. 855-861.

—G. Stoops, and C. Sys, 1977, The micromorphology of gibbsite forms in soils: Soils Science of America, v. 28, p. 136-143.

Frank, J.R., 1981, Dedolomitization in the Taum Sauk Limestone (Upper Cambrian), Southeast Missouri: Journal of Sedimentary Petrology, v. 51, p. 7-17.

Gilkes, R.J., A. Suddhiprakarn, and T.M. Armitage, 1980, Scanning electron microscope morphology of deeply weathered granite: Clays and Clay Minerals, v. 28, no. 1, p. 29-34.

Ixer, R.A., P. Turner, and B. Waugh, 1979, Authigenic iron and titanium oxides in Triassic redbeds (St. Bees Sandstone), Cumbria, northern England: Geological Journal, v. 14, p. 179-192.

Lonsdale, P., V.M. Burns, and M. Fisk, 1980, Nodules of hydrothermal birnessite in the caldera of a young seamount: Journal of Geology, v. 88, p. 611-618.

Scholle, P.A., 1979, A color illustrated guide to the constituents, textures, cements, and porosities of sandstones and associated rocks: AAPG Memoir 28, 201 p.

Spiro, B., and I. Rozenson, 1980, Distribution of iron species in some "oil shales" of the Judea desert, Israel: Chemical Geology, v. 28, p. 41-54.

Walker, T.R., E.E. Larson, and R.P. Hoblitt, 1981, Nature and origin of hematite in the Moenkopi Formation (Triassic), Colorado Plateau - a contribution to the origin of magnetism in redbeds: Journal of Geophysical Research, v. 86, p. 317-333.

—B. Waugh, and A.J. Crone, 1978, Diagenesis of first-cycle desert alluvium of Cenozoic age, southwestern U.S. and northwestern Mexico: Geological Society of America Bulletin, v. 89, p. 19-32.

Recommended References: Miscellaneous (Wood)

Meylan, B.A., and B.G. Butterfield, 1972, Three-dimensional structure of wood: London, Chapman and Hall, Ltd., 80 p.

Stein, C.L., 1982, Silica recrystallization in petrified wood: Journal of Sedimentary Petrology, v. 52, p. 1277-1284.